AF700039

Calendering & Winding Technology

For: Ray Bender

Rec'd July 23, 1985

Cathy S. Forbes
Editor-in-Chief
TAPPI PRESS

Technical Association of the Pulp and Paper Industry
Technology Park / Atlanta
P. O. Box 105113
Atlanta, GA 30348
404/446-1400 • TWX 810-757-0145

TAPPI PRESS

Technology Park/Atlanta, P.O. Box 105113, Atlanta, GA 30348

Printed in the United States of America

International Standard book Number 0-89852-427-X

Preface

This TAPPI PRESS report is a compilation of twenty-three technical papers presented at the TAPPI 1985 "Calendering Technology for Improved Printability and Winding Techniques for Better Roll Structure" Seminar held in Escanaba, Michigan.

TAPPI wishes to thank Thomas S. Greiner who served as Seminar Chairman and compiled the papers in this report.

Contents

Contributors

Jack Beach, Mead Corp., Publishing Paper Division, Escanaba, MI

Ray S. Bender, Appleton Machine Co. Appleton, WI

Morris R. Curry, Publishing Paper Division Escanaba, MI

E. Russell Dreschsel, Madison Paper Industries, Box 129, Madison, ME

Gary F. Fecteau, Appleton Machine Company, Appleton, WI

Theo Fernholz, Jagenberg AG, Duesseldorf, West Germany

G. Forsberg, Consultant, Beaverton, OR

Robert L. Grant, Beloit Wheeler Corp., P.O. Box 2062, Kalamazoo, MI

Thomas S. Greiner, Jagenberg, 645 Madison Avenue, New York, NY

George Grenon, Hewitt Machine Co., P.O. Box 688, Neenah, WI

Steven S. Hardy, Boise Cascade Corp. North Coated Division, Runsford, ME

David W. Johnson, Mead Corp., Publishing Paper Division, Escanaba, MI

Dick Johnson, Mead Corp., Publishing Paper Division, Escanaba, MI

Glenn N. Johnson, Midtec Paper Corp., Kimberly, WI

John Kilmartin, Consultant, 1757 Greenbriar Drive, Kalamazoo, MI 49008

Pekka Komulainen, Wartsila Jarvenpaaworks, SF-04400 Jarvenpaa, Finland

Erhardt E. Munch, Kleinewefers, Krefeld, West Germany

Jurgen Schlunke, Kleinewefers, Krefeld, West Germany

Wimmar Schmitz, Kleinewefers, Krefeld, West Germany

David Wereley, Midtec Paper Corp. Kimberly, WI

Earl Winchester, Mead Corp., Publishing Paper Division, Escanaba, MI

One
Calendering and Supercalendering—Theory and Practice

ABSTRACT

Theory

Understanding and applying the theory of nip mechanics is an essential replacement for the traditional method of "trial-and-error".

Early concepts are reviewed and rejected. The current best theory is that paper surfaces tend to "replicate", or mold to the surface textures of the rolls in a nip.

PRESSURE, HEAT, and TIME are the factors involved in replication, and they can be used in a variety of ways. However, there are only certain changes that can be made to paper in a calender or in a supercalender nip.

Nature of the filled rolls is reviewed, and the manufacturing method is briefly outlined.

Practice

Specific cases are described in which an understanding of supercalendering theory could have saved time and money.

Proper application of the principles opens the door for use of mark-resistant synthetic roll covers and for generating final finish on the paper surfaces at the machine calender.

CALENDERING AND SUPERCALENDERING - THEORY AND PRACTICE

Introduction

In a number of talks prepared for technical sessions of the TAPPI Calendering Committee and for the TAPPI Short Course last year, the emphasis has been on both the theory of what happens in a supercalender and some practical examples of how production experience really does conform to the theory. Now, the relationship between theory and practice will be even more specific.

The purpose of technical sessions, short courses and seminars is to present something that you can take back to your company that will have an economic payout. It is fine to have a healthy personal curiosity about these matters. That has been my reason for continued thought and study. However, I was never in a situation where the things I observed and learned could have a direct application to a production situation. I do hope that some of the people who have listened to me were able to effect logical changes and economies in their mills.

We must look at the theory first when we discuss theory and practice, even though the theory could not have evolved without some practical observations of what normally happens in the real world.

PAST AND PRESENT VIEWS

When we talk of calendering and supercalendering, we refer to it as an art rather than a science. The operating people who were most helpful, and indeed, essential to good performance had learned what to do through "trial-and-error". When a certain problem appeared on the supercalender reel or on the finished shipping rolls coming off a winder, they related it to previous experience. They had learned, through trial-and-error, what changes could be made to minimize or remove the problem. This expertise required a sincere interest in making good shipping rolls, and and it required many years of this interest, so that repetitive problems could be recognized and corrected quickly. The action taken may have been partly right and partly wrong. In any case, mills survived and the industry grew with the experienced old-timers putting out fires.

We now see a new attitude in workers that requires a change in the way the finishing department operates. Many of the young workers lack the interest in improving performance and they don't observe, and think about, the reasons for poor quality in the reel or shipping rolls. There are certainly many of our young people who still are well motivated, but these reach out for higher education and are not normally available to the mill labor force.

CHANGE IN PROCEDURES

There is more to the problem than just attitude. The paper machines and the finishing machines are bigger and faster. Even the old-timers in their best days would have had a hard time learning and applying the expertise under present day production rates.

There had to be an answer, and we see it in the ever-improving instrumentation. Instruments have been designed to sense and display rates of flow, pressures, basis weight, moisture content and a variety of other physical factors that are involved in the making and handling of paper. These instruments replaced the seat-of-the-pants judgments of the operator. In the early days, he was still required to make the actual adjustment on the machine. The instruments were then further improved to perform the mechanical adjustments as well as to sense and indicate a problem. The results have been good, and we see extensive instrument control in the entire papermaking process.

WHY IS CALENDERING DIFFERENT?

The main problem with using instruments to control the performance of a calender, and particularly a supercalender, is that we have not really understood what happens to the paper when we pass it through these nips, nor have we understood the limited number of improvements that can be made by simple mechanical adjustments.

An instrument man was quite concerned last year when I said to an assembled group that we must be careful not to indicate that selective pressure can be applied to a supercalender to correct a variety of hard and soft spots across a reel. He knew that they could sense the problem of caliper variations, and he assumed that corrective action could be at the direction of an effective instrument system. The simple fact, is that only certain things can be done in these rolling nips, and we must understand this to do away with the old assumption that bad paper can be changed into good paper in a supercalender nip.

GENERAL DESCRIPTION

Calendering is most simply described as "pressing in a rolling nip". The purpose is to generate a smooth, level surface, so that the paper will be proper for printing or for some other converting operation. For many years, a suitable surface finish was possible by using the chilled iron rolls in a machine calender. In some cases, the surface was further improved by applying water or starch during the calendering operation. However, the demand for a better surface led to the off-machine operation that we in the paper business logically call supercalendering.

It is impossible to change the surface texture of paper in a production situation without densifying the entire thickness. There has been one successful research effort to change only the surfaces of a newsprint sheet by applying very high temperature in a sharp, quick nip. The interior of the sheet does not have the time to get hot, so it springs back to a bulky formation. This phenomenon is beyond reach in a conventional calender, but it does support the theory that we are discussing here today.

Machine calenders will be discussed later in the program, so our emphasis will be on supercalenders.

SUPERCALENDER

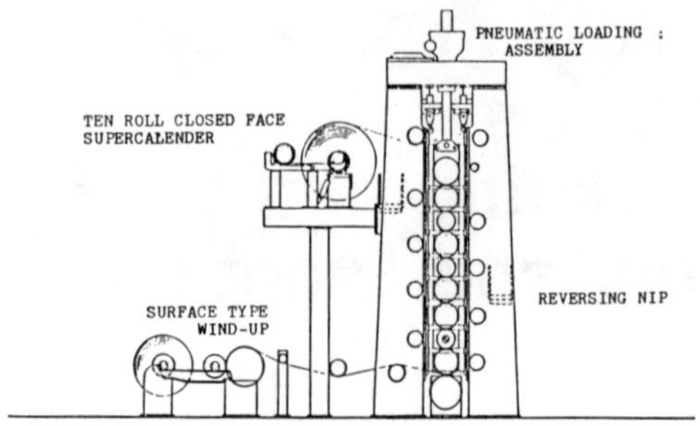

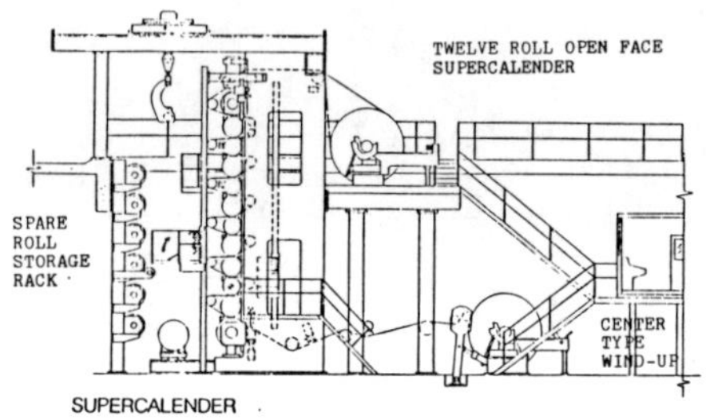

SUPERCALENDER

Figure 1. Supercalenders
10 Roll "A Frame" - Closed Face
10 Roll Open Face with spare roll rack

Two general configurations are shown. Both have resilient filled rolls alternating with metal rolls.

The metal rolls are most generally made from chilled iron for maximum surface hardness. The closed stack shows ten rolls with a reversing nip. The open-face stack shows twelve rolls with a reversing nip and a spare roll storage rack. We will talk about the reversing nip later.

In an effort to cover a lot of ground in the alloted time, brief statements will be made about many aspects of both the equipment and the process. These can be expanded and explained in the discussion sessions.

Most supercalenders are in a vertical arrangement.

Most supercalenders have a single drive input. The historic plan was to drive the bottom roll. Torque was transmitted upward with each roll driving the one above it by surface contact. Very little surface speed is lost between the bottom roll and the top roll. A much greater speed difference exists between the surface of each filled roll, and the adjacent chilled iron rolls. We will discuss this later.

Most supercalenders use many nips. Some use a reversing nip, where two filled rolls run together. This presents each side of the paper to the iron roll during part of the web run. Others operate without the reversing nip, and only one side of the paper is pressed against the iron rolls.

Supercalenders use hydraulic or pneumatic systems to add to the weight of the rolls. This added pressure is adjustable.

Intermediate rolls, including all the filled rolls, are straight-faced with the same diameter from one end to the other.

The bottom roll, frequently called the king roll, must be crowned with a larger diameter in the middle to offset the weight of all the rolls, and to adjust for the external loading; or it must be of a design that will use an outer shell of constant diameter and compensate for the weight by taking the deflection in a large fixed center beam. We call this a crown compensating roll, and there are several designs in use on large supercalenders. When the crown compensating rolls came into use, the drive input was moved up to the lowest of the small intermediate iron rolls.

Some supercalenders use a surface type wind-up, and others wind the paper by applying torque to a large shaft, or spool, in a fixed position wind-up stand.

EARLY CONCEPT

No one seemed to question that a machine calender consisting of all chilled iron rolls caused densification and generated an improved surface finish by pressure alone. However, when a much better surface finish resulted from the nips between chilled iron rolls and filled rolls, everyone assumed that some rubbing action or slippage took place. Most people felt that a change of path of the larger diameter filled roll surface caused the paper to rub against the smooth surface of the smaller chilled iron roll.

ROLLS IN A SUPERCALENDER

WHY IS THE FILLED ROLL BIGGER?

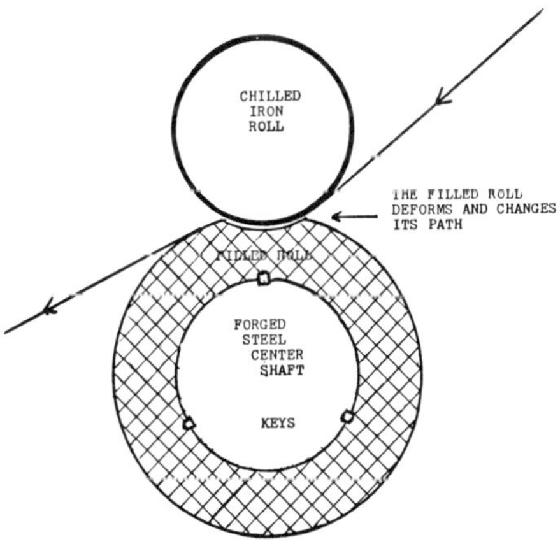

FILLED ROLL IS BIGGER THAN THE CHILLED IRON INTERMEDIATE IRON ROLLS BECAUSE THE CENTER SHAFT MUST BE LARGE ENOUGH TO RETAIN THE LOCKED-IN PRESSURE AND TO RUN AT OPERATING SPEED

Figure 2. Paper in the nip.

It was established some years ago that the surface speed of the filled roll outside the nip is actually faster than the surface speed of the chilled iron roll that is driving it and of the one that it is driving.

RELATIVE SURFACE SPEEDS

SUPERCALENDER ROLLS

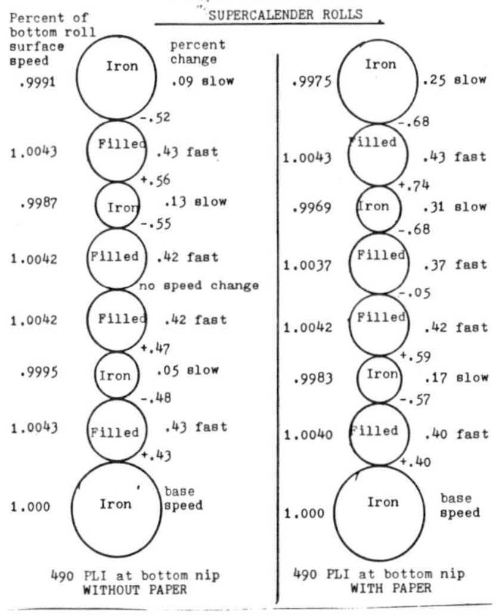

NOTE THAT SURFACE SPEED DIFFERENTIAL BETWEEN THE BOTTOM ROLL AND THE TOP ROLL IS GREATER WHEN PAPER IS BEING RUN.

Figure 3. Surface Speed

Figure 3 shows the relative surface speeds in an eight roll laboratory supercalender. Two things are particularly noteworthy. First, that the total speed loss from the bottom to the top is less than the speed difference between each iron and filled roll. The second thing to note is that the speed loss at the top is greater when paper is running. This does indicate that there is some shear within the paper; but for all practical purposes, we consider the surface speeds to be matched in the high pressure part of the nip. This means that the filled roll surface must slow down as it goes into the pressure part of the nip and must speed up as it comes out.

When some mills dropped the reversing nip and still got good finish on both sides of two-side coated paper, they shot down the "slipping and polishing" theory. Some of the die-hards called attention to the speed change going in and coming out of each nip, and they said "that must be where the slipping takes place". Dave Pfeiffer, about fifteen years ago, said there is no significant slippage between the paper and the roll surfaces. He demonstrated it quite well in a paper that was finally published as part of the proceedings of the TAPPI Finishing Conference held in Hershey, Pennsylvania in the fall of 1980. Dave said that the "micro-slip" areas were of too short duration, and at less than functional pressure. He concluded that the finishing action results from the tendency of the fibers and coating to replicate the surface texture of both rolls in a pressure nip. I must stress that the paper only tends to get as smooth as the roll surfaces. It cannot get smoother, but it can, and generally does, fall short of 100% replication. When we understand and accept this theory, we find that most of the production problems can be explained and can be corrected. You will hear me say loud and clear that many of the corrections called for are not possible in the supercalender, and must be taken care of elsewhere.

HOW DO WE GET MAXIMUM REPLICATION?

If a material flows easily, it will mold to a surface texture quickly and fully. My best example, and one with a practical use, is "silly putty", which I call "Super Surface Texture Tester". Silly putty will replicate a surface in fine detail very quickly with light pressure. If you check its surface quickly against reflected light, you will see how smooth the tested surface really is. Silly putty can be used to match a roll surface against a paper sample of known gloss by pressing against the paper edge and the roll surface at the same time. Fine detail will fade quickly, but this simple test can tell a lathe operator whether he has sanded a plastic roll surface enough, so that it will be suitable for use in a calender.

When we talk about replication, we talk about flow in the coating and in the paper substrate. The material must flow to replicate the roll surfaces. Unfortunately, the paper and the coating that is supercalendered does not always flow enough to result in the desired gloss and smoothness. Even more significant is the fact that variations across the web result in variations in the amount of replication. The existence of these variations is what separates practice from theory.

Let's look at the factors that contribute to replication. We work with PRESSURE, HEAT and TIME.

SUPERCALENDERING
Factors Contributing to Replication

Pressure	- Generally adjustable (Varies with temperature change)
Heat	- Function of pressure and speed Not normally controllable
Time	- Function of: Number of nips Fixed Machine speed Adjustable Width of nips Fixed as to roll hardness Varies with heat in rolls

Figure 4. Factors Required for Replication

PRESSURE is under operator control. It is a readily variable factor, as far as gloss generation goes. However, its use has some limitations. In a supercalender with a fixed crown in the bottom roll, there is only one operating pressure that results in a level nip. The operator must run at, or near, this pressure to prevent hard or soft ends on the reel.

The fairly recent use of crown compensating rolls allows a much greater flexibility in selecting a pressure level because the system permits even pressure from one end to the other, regardless of what level of nip pressure gives the best results. However, we will see that variations in pressure will result from temperature variations across the roll face.

HEAT in a supercalender has historically been from generated temperature in the surface of the filled rolls. Heat results from hysteresis action each time the roll surface is compressed and relieved as pressure is applied in the nips. It takes a long time for this heat to build up in a filled roll body and to transfer to the chilled iron rolls. The level of temperature is a direct function of the amount of pressure applied and the speed of operation. The faster the stack runs, the more heat is generated.

Two major changes in supercalender design in recent years has dramatically improved the use of heat. In the first case, we find that the warm oil used inside the crown compensating rolls supplied heat to the large bottom and top rolls. The second change is more directly related to the use of heat. Hot water or hot oil is passed through specially designed intermediate iron rolls. This temperature can be controlled and the use of these rolls has liberated the filled rolls from the burden of supplying the uncontrolled heat.

TIME is an interesting part of supercalendering. Each cotton filled roll deforms when it passes through the nip, so there is some width to the pressure area. This compares to a line contact between two iron rolls in a machine calender. The width of the nip means that the paper is under pressure for some period of time. The total number of nips will multiply the time factor. Speed of running will also determine how long the paper is actually under pressure in a supercalender. Another variable is the fact that a hot cotton roll will deform more in the nip than the same roll when it is cold. Remember this when we talk about pressure variations across the stack in production situations.

You can easily understand that since the factors contributing to replication are identified there can be many ways in which they can be balanced against each other to get the best result.

LABORATORY TESTING AND PRODUCTION CONDITIONS

It is easy to determine how much gloss and smoothness result in small samples or in narrow webs in the laboratory. The amount of coating, the coating formulation, the moisture in the base sheet and other factors can be measured and controlled. Such testing can certainly indicate the requirements for a production run. Unfortunately, the inconsistent quality of paper produced does not necessarily relate to the laboratory tests. Many variables exist in production. We must understand what the variations are, how they happen, and what effect they have on the product, to clearly understand supercalendering. We said earlier that experienced operators learned what action might help after the objectionable results showed themselves. I am saying now that we can anticipate the bad result and take corrective action before the supercalender produces bad paper.

Let's look at the least stable element in a supercalender.

FILLED ROLLS

The heart of a supercalender and the thing that makes it different from a machine calender is the filled rolls. These rolls will deform under pressure in the nips, but will still be firm enough to force the paper and coating to flow, and to replicate the roll surfaces. Filled rolls are made almost exclusively from 100% cotton fibers. Rolls used in Europe and embosser rolls have some wool fibers mixed with the cotton. These fibers retain the locked-in pressure, and the roll surface assumes a smooth texture relatively the same as the surface of the opposed chilled iron roll. Hard filled rolls glaze up more quickly during a run-in period, and they reach a higher level of surface gloss.

The filled rolls are built by placing discs of material over the upper end of a solid forged steel shaft standing on end in a large hydraulic press.

Figure 5. Building a Filled Roll

The lower journal extends down into a cavity in the ram, or piston, of the press which is normally located below floor level. The lower nut and retaining, or pressing, ring rest on the ram itself. When a portion of the discs have been manually placed over the top of the shaft, they are moved down to the bottom by hand and a new pressing ring is positioned on top of the pile. The press head is lowered or spacers are swung into place to resist the upward thrust of the hydraulic ram. The shaft is forced upward through the discs, and they are compressed into a firm solid mass of cotton fibers. This section of fill will remain under pressure for as much as two days. The same operation is repeated three or four or more times until the shaft is full of compacted discs. At this point, the upper nut is put in place. When the ram lowers this last time the shaft feels the full force of the compressed material. The pressure applied and locked into a filled roll is between 8,000 and 12,000 pounds per inch of fill area. This means that the new large rolls that you will see in the mill have more than five million pounds of pressure locked in to keep the roll firm during its useful lifetime. The lifetime of each roll refill will vary greatly based on how often the sheet breaks while running through the stack, or on how badly the roll shape becomes distorted by running bad paper or by less than thoughtful operation of the supercalender itself.

The filled rolls will mark easily from wrinkles or wads of paper that pass through the nips. When marks are severe enough to show on the paper surface, a roll must be removed from the supercalender and replaced by a truly round and smooth roll. Another reason for removing a roll for resurfacing is to remove the distorted shape caused by running paper with a bad profile or by poor operation, as already mentioned. After a roll has been used, damaged and refinished a number of times, it has reached a minimum diameter. The roll is then sent out to be stripped and refilled to the proper maximum diameter.

HARDNESS OF FILLED ROLLS

We refer to all filled rolls as the soft rolls in a supercalender, because they are softer than the chilled iron rolls. Filled rolls can be made to a specific hardness by selecting the proper material. The total range varies from about 82 to 92 on the Shore "D" scale. The Shore hardness tester was developed to measure the displacement of a rubber roll surface by pressing the rounded head of a spring-loaded pin against the roll. For lack of anything else to use, the Shore instrument was modified by sharpening the spring-loaded pin and calibrating the spring to read the very small amount of penetration into the surface of the compacted cotton discs. This Shore "D" instrument is not a very good tool, but it is all we have. The softer rolls are made by using bulky discs made from long cotton waste fibers. The harder rolls use the same type of waste cuttings from the textile industry, but the fibers are refined more, and the webs are formed on fourdrinier paper machines at very slow speed. It is interesting that the soft rolls receive very high pressure in manufacture, and the harder rolls are properly firm with less locked-in pressure. The most widely used material in the United States and Canada is a firm sheet, like thin cardboard, and is usually pressed to a hardness of 88 to 90 Shore "D".

USE OF ROLLS IN THE MILL

Soft rolls deform more in the nip. Pressure is low due to the width of the nip, more heat is generated and the rolls can be sanded with relative ease to remove surface marks. These are most commonly used in narrow, fractional deckle supercalenders. Various widths of paper are run on these stacks. Normal good practice calls for running the wide rolls first and then progressively narrower ones. This permits sanding away, at one time, all the edge marks that have been pressed into the roll surface.

Full width supercalenders usually use hard rolls, and the mills learned years ago that it is imprudent to sand these. They are too big, too hard, and sanding rolls to relieve a hot ring or correct a hard part in the reel has only limited benefit for a short time. Most mills replace the rolls when there is a problem, but a few still try to chase away pressure problems by sanding these big rolls for shape. Of course, there are some specific cases where a little sanding can be effective, so the art is not totally lost. Ask your production people about this during the panel discussions.

PRACTICAL ASPECTS OF SUPERCALENDERING

Now, we look at the important part - What does all this mean?

We know that we are going to apply pressure to the web of paper in several nips. Heat will be present, and we want to run at a good productive speed.

We desire even replication from one edge to the other, and we want even densification, so that residual caliper will be the same all the way across the sheet. Replication will be judged by measuring gloss and smoothness. Usually they go together, so common practice is to monitor gloss while the supercalender is running. This will be verified in the lab from samples taken near the end of each reel. If the gloss reading is low, the operator can increase pressure, increase temperature or slow down the stack. None of these is a good idea, if the operating parameters are in a range that normally yields a good result. The best plan is to change the coating formulation slightly to get more replication. Moisture level in the paper is very significant, but we will cover this when we talk about the mechanical condition of the reel, rather than surface characteristics.

One tool that is used to get good gloss on coated paper is the application of steam at some of the ingoing nips. The moisture and/or the temperature helps the coating surface to replicate. Too much steam will cause the coating to pick off and stick to the hot roll surfaces.

MECHANICAL CONDITION OF THE REEL IS MOST IMPORTANT

Much of the discussion in winding sessions is related to hard ends, soft ends, corrugation and other problems that show up on the shipping rolls. If these objectionable variations are present, the roll has "poor mechanical condition". It may be improved by rewinding, but it is certainly better to have good mechanical condition in the first place. The problem most generally relates to caliper variation across the web. In supercalendered paper, the caliper variations don't show up until the paper has been subjected to the high pressure nips. A reel of paper before supercalendering may have a level hardness profile. This means that the caliper is really quite even and the reel looks good. The same reel may have a very poor hardness profile at the supercalender wind-up.

There are two possible reasons for a reel that looks good to show up as a bad reel after supercalendering. The fault can be that the paper densifies differently at certain places even though equal pressure, temperature and time are applied across the supercalender. The second reason can be that the filled rolls are distorted and that variations in pressure, temperature and nip width are resulting in uneven densification. It is hard to separate the two.

I asked many good supercalender people, over the years, how a known moist streak in the paper shows up on the reel after supercalendering. Some said the reel will always have a hard ring where the paper is more moist. Others said the spot would always be soft. Still others said "sometimes hard and sometimes soft". Actually, they were all stating true conditions, and the difference is in the type of paper with which each person was familiar. A lightweight coated sheet will always have a hard spot where the paper is moist because the supercalender will be cooled by the moisture. The iron rolls will become smaller in diameter and less pressure will be applied to the web. The same condition in a seventy or eighty pound sheet will show up as a soft spot in the reel because the moist fibers densify more readily. There may be some tendency to cool the supercalender rolls, but the densification of the fibers has a greater effect. In the lightweight sheet, there are fewer fibers and the caliper reduction is less significant than the change in pressure due to cooling. The group that had no strong feelings about the moist streak were used to supercalendering papers where the opposed reactions would be almost in balance.

Another possible reason for hard parts in the supercalendered reel of coated paper is that a moist streak will pick up more coating. The coating does not densify like fibers, so the caliper will be high. I don't think this is valid, but it does help point out that it is not easy to know the real cause and the logical correction when problems occur.

WHAT CAUSES MOST SUPERCALENDER PROBLEMS?

Problems can be caused by uneven pressure applied to the web, or by uneven response when the pressure is constant from one edge of the web to the other.

I would like to cite a specific example and then we will get into the meat of the practical side of supercalendering. Several years ago, when the dean of supercalender men retired, his mill policy dictated that the paper machine people would be responsible for the two one-hundred sixty inch wide coated paper supercalenders. When a problem such as a corrugation or hard rings would appear on the supercalender wind-up, the new bosses looked at the mysterious stack of rolls and had no idea what to do to correct the problem. Happily, they felt that their paper making experience would permit them to help out by making some change on the paper machine. They tried it, and it worked. These people soon reached the conclusion that if the filled rolls were in good shape, and if temperature was level across the stack, any defect that showed up on the supercalender reel is a paper machine problem, and should be corrected at the paper machine.

Let's skip over the problem of uneven pressure in the supercalender causing poor reel condition. Most operations have two supercalenders taking paper from one machine. All the paper goes through one coater, whether it is part of the paper machine or an off-machine coater. Both supercalenders should have the same problem, if the defects are in the paper. Any good supercalender man will confirm that the problem is in the paper by swinging one reel end-for-end to see if the problem at the reel also moves. Then, he will call the problem to the attention of the paper machine operators.

Unfortunately, there are many times when the supercalender people can point out the problem, but they don't have any constructive suggestions how to correct it. Understanding the theory of supercalendering is the only way to know what changes are logical and to have some idea of how much corrective action is required.

MOISTURE IS OUR MAIN PROBLEM

When poor reel condition results from uneven densification of the paper in good level nips, the problem is either basis weight variations or moisture variations. I think we can rule out basis weight variations in general, because the instruments are well refined and papermakers believe what the instruments say. It was not always so. It is not too many years since papermakers were convinced that a straight dry line across the wire meant even distribution of fibers out of the head box. Incidentally, I don't believe you can have a streak of high basis weight without a corresponding streak of high moisture that carries right through to the reel. On the other hand, you can have a very level basis weight profile and a horrible moisture profile because a whole nest of things can vary across the machine in the drainage, pressing and drying of paper.

We have already said that a moist streak can have directly opposite effects based on the type of paper. Here is another example that occurred in two different mills. The mills were in different parts of the country, producing very different grades of paper, and the problem was one that all papermakers are well familiar with. The problem was dry edges, and a moist middle. We could talk all day about dry edges as being characteristic of all paper machines. However, in both cases the problem was of such a magnitude that both mills blamed the supercalender, and particularly the filled rolls,for not being able to knock down hard ends on the reel. Each of these mills was trying to run non-coated paper with a high moisture level. In my limited experience, I believe that paper from most paper machines that has an 8% average moisture will probably be as low as 4% at the edges, and as high as 11% in the middle. The moist middle will densify easily at low pressure and will replicate nicely to a good gloss level. The dry edges, on the other hand, resist both densification and replication. If a supercalender can be operated with negative crown it still cannot press the edges hard enough to reach the densified caliper that comes so easy in the moist middle. Extreme pressure at the roll ends may help for an hour or two, but the filled rolls gradually get more compacted at the ends and the crown shape no longer applies. Both mills were finally able to operate by bringing the moisture level down and/or leveling out the moisture difference by overcrowning the press section.

MORE SPECIFICS ABOUT MOISTURE

An obvious answer to the desire for a high moisture content all the way across the web is to remoisten the sheet selectively after it has been partially or completely dried. This is practical, because the need for additional moisture is in a constant pattern, and in known areas. It is common good practice in mills supplying supercalendered groundwood paper.

Moist streaks are a whole different ball game. They can appear any place across the machine, and come from worn wires or fabrics or worn water removal devices in the drainage section. They can come from plugged cleaning showers, worn felts, old rubber covers on press rolls. They can come, although not as often as has been suspected, from varying temperature across some dryers.

Whatever the cause, a bad moisture profile is the most significant and perhaps the last important misunderstood problem for supercalendering. In defense of the people who run paper machines, I will say that even the best run machine is subject to a gradual degradation of moisture profile. It is the nature of wires and felts to wear and to plug up. The key is for the machine people and mill management to know at what point the degradation begins to cost a lot of extra money in the finishing department.

HOW UNDERSTANDING THEORY CAN HELP

To wind up the idea that it is good to understand the things I have reviewed, I'll touch on two cases with which I am vaguely familiar. I wish I knew more details, but my limited knowledge will still make a valid point.

The first case is one in which a mill intended to raise the moisture level in supercalendered speciality ground-wood paper. They knew that selective addition of moisture would be necessary and they planned accordingly. What was not planned was the fact that the moist web would behave very differently in the supercalender as compared to the dry sheet that had been run for years. It is only in hindsight that I say this, but this mill could have anticipated much more elongation in the top nip, so the paper would be slack and tend to show serious soft wrinkles. This is the same thing that happens in a reversing nip. The nip tends to put out more paper than the next nip takes in. Anticipating this problem would have called for a fly roll with a deep pattern of ridges after the first nip.

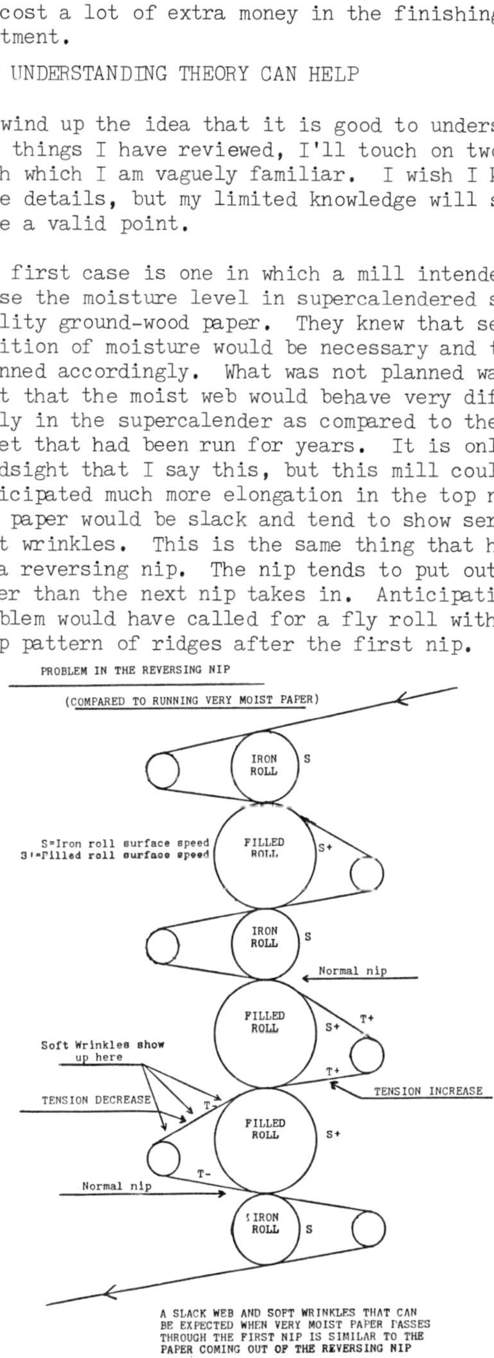

Figure 6. Reversing Nip

The higher moisture level would permit proper gloss and smoothness, so the normal pressure used for the previous dry paper was no longer necessary. A proper plan would have anticipated lower nip pressure. This might have called for negative crown in a spare intermediate iron roll or temporary negative crown in one or more filled rolls.

Another problem was that moist streaks in an already moist sheet became very critical. These streaks elongate even more than the rest of the web, so a bubble builds up at the ingoing nip. The bubble will finally fold on itself and cause a short cross-machine calender cut.

Other problems could also have been anticipated, although they may not be as critical as the extreme densification and elongation of the sheet. We can be sure that a mill that had adjusted over the years to dry edges and a less dry middle would also have problems in this area.

The second case where understanding of supercalender theory can be helpful is in fitting existing supercalenders to the unusual characteristics of a new type of paper. It is my understanding that at least one producer of lightweight coated paper could not successfully run a European-type coated paper with special fillers and coating on their existing large supercalenders. Serious consideration was given to replacing the existing equipment with something similar to the supercalenders that were used for the special paper in Europe. I don't know what was decided because I don't have constant contact with the mills anymore. Anyway, I suggest that proper modification can be planned for existing equipment, if the new type paper is evaluated regarding replication under pressure with heat and time available.

Along these same lines, a paper outlined the history of a European mill that bought a third large supercalender to help two others finish the paper from one paper machine. Maybe they were correct in determining that it was not possible to obtain the finished characteristics that were required at a speed that would allow two stacks to key up. I may be wrong, but there must have been a less expensive way.

PLASTIC ROLL COVERS

I must not leave the subject of supercalendering without saying again what I have said several times before. Mark resistant synthetic roll covers have been used successfully in supercalenders for several years. The most successful has been the Supertex roll, which is a combination of synthetic fiber mat and epoxy resin. Many covers were applied to cotton roll bodies, but it was determined in 1982 that this was not acceptable for use at normal supercalendering pressure. The practice of overloading the crown to densify dry paper edges causes very high nip pressure at the ends of the rolls. This generated heat in the cotton roll body, and destroyed the ends of the one-half inch thick plastic cover. The cover itself was strong enough to take a normal amount of this extra end pressure, so Supertex covers on steel bodies can operate successfully in conventional supercalenders. Unfortunately, something was changed in the manufacturing procedures, and quite a few of these rolls have been unsuccessful. There had just been a similar situation when I gave a paper at the TAPPI Finishing Conference in Wausau in 1978. I'm sure these covers will get back on track soon. In the meantime, the successes did prove that the industry is close to the roll that has been dreamed about for more than twenty years.

This opens the door to more and more finishing on machine calenders as a replacement for some of the work done today on supercalenders.

REFERENCE LITERATURE

1. Pfeiffer, J.D., "The Finishing Action of a Supercalender Nip", Presented at the Fourth TAPPI Finishing Conference, Portland, Maine, October 1970.

2. Pfeiffer, J.D., "Supercalendering - 1980 Update", TAPPI 1980 Finishing and Converting Conference Proceedings, pp. 19-31.

3. Peel, J.D., Hudson, F.L., "Application of Elastic Theory to a Supercalender Nip", Paper Technology (Technical Section) B.P. & B.M.A., London, Vol. 50, No. 7, July 1967.

4. Kilmartin, J.A., "Supercalendering - What is is?", TAPPI 1979 Annual Meeting preprints, Feb. 1979, pp. 215-221.

5. O'Gara, W.J., Wilberg, W., Schiller, G., Aschenbrenner, W., Lazar, J., "Some Super calender Nip Phenomena", TAPPI, Vo. 49, No. 9, Sept. 1967, pp. 422-424.

6. Hunger, G.K., "The Influence of Calendering on Sheet and Coating Densification and on Final Printability", TAPPI, Vol. 50, No. 7, July 1967, pp. 372-379.

7. Kilmartin, J.A., "Final Report - CA 10035 'Ongoing Survey - Supercalender Operations'", Presented at 1979 TAPPI Finishing and Converting Conference.

8. Peel, J.D., Kerekes, R.J., Baumgarten, H.L., "Calendering and Supercalendering - An Appraisal of Current Practice, Understanding and Needs", Introductory paper for a Symposium on Calendering and Supercalendering of Paper, which was held at University of Manchester Institute of Science and Technology, Manchester, England, on 1st and 2nd September, 1975.

9. Crotogino, R.H., "Temperature Gradient Calendering", TAPPI Journal, October 1982, pp. 97-101.

10. Atkins, J.W., Rodencal, T.E., Vickery, D.E., "Correcting a Poor Moisture Profile Improves Productivity", TAPPI, Vol. 65, No. 2, Feb. 1982, pp. 49-52.

Two On-Machine Soft Calenders

G. Grenon

Will on-machine supercalendering be added to the continuous papermaking process?

There's a very real possibility that it will. And if you decide that you'd like to give serious consideration to the costs and the benefits, two broad categories of "on-machine soft calenders" are being offered in today's marketplace. One is from the U.S., the other from Europe.

The goal of both types of processes is to produce a sheet suitable for fine offset printing. One which offers good ink holdout, yet does not have the glare associated with high gloss papers, since dull, matte or velvet finishes are becoming more and more popular all the time.

Preservation of bulk is an important consideration, of course. Using a supercalendering process to obtain high finish invariably results in a densifying of the sheet, something papermakers would like to avoid, if possible.

A couple of additional factors which tend to favor lower finishes in the printing grades of the future are these:

1. Basis weights are being reduced to offset rising mailing costs and the cost of fiber.

2. A lower finish preserves opacity and makes it less necessary to use costly additives.

Still in all, it is essential to produce a sheet having a very uniform caliper profile and level surface in order to assure quality printing performance without making use of the crutch of excessive calendering.

Now, there's a balancing act for you!

Up to now, there have been two general approaches to the problem of how to produce coated papers with low finish.

Matte coated papers nearly always carry a smaller amount of coating than enamel papers and receive little or no calendering. They are not finished on supercalenders or other off-machine equipment.

Dull coated papers receive the same amount of coating as their enamel counterparts and are finished on modified supercalenders or on calenders with etched rolls, tandem calenders or other special equipment.

The objective remains the same---to produce a level sheet with the same or better ink receptivity as a supercalendered sheet, and to do this with as little loss of bulk as possible.

The traditional methods for producing low finish paper grades has typically produced a sheet with inferior printing properties because of less coating and little, if any, surface treatment. Or else it has been necessary to choose secondary finishing with its increased production costs, higher broke losses and paper deterioration due to the extra unwind-rewind cycle.

So much for history. Now let's consider the potential of the two so-called "new approaches".

First, let's take a look at the system known as "Mat-On-Line," developed by E. Kuesters in Krefeld, West Germany.

"Mat-On-Line" makes use of a smooth surface, variable crown type roll operating with a plastic-covered roll. A roll cover of proprietary composition is used in single or double nip machine calenders in conjunction with a "Vario Swimming Roll." This type of supercalender imparts a superior smooth, but not glossy, finish through the use of comparatively low nip pressures and moderate temperatures.

Though nip pressures of more than 1000 pli can be tolerated by the special roll cover, installations tested with numerous grades suggests that considerably lower pressures are best, producing the desired surface enhancement without appreciable reduction in bulk.

The main thing seems to be the need for uniform conditions in the nip. A normal crown compensating roll is not always satisfactory, especially with wide paper machine applications. The manufacturers believe that, in order to compensate for minute variations in cover wear and also profile characteristics of the sheet, either the mat-on-line roll or its mating roll should be a Vario Swimming Roll.

Here's what the Vario Swimming Roll looks like in schematic. (Slide #1). It consists of two opposing pressure chambers with differential pressures adjustable as needed. The Vario elements---goggle-shaped twin enclosures---are mounted on the stationary roll shaft in opposing pairs, creating separate enclosures within the two chambers.

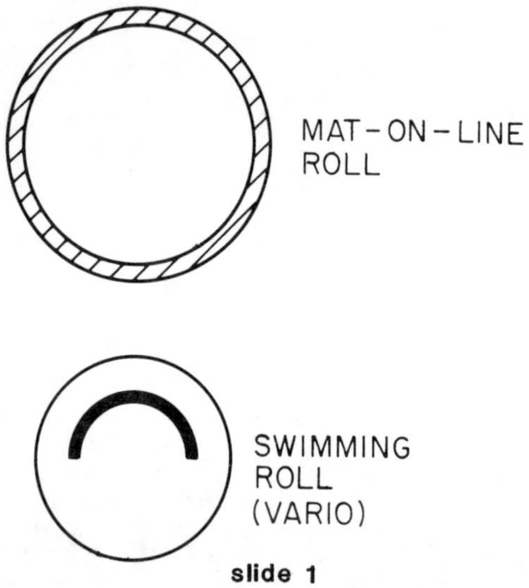

slide 1

Mounted at critical points, these elements permit small area corrections of nip pressure as needed, through changes in oil pressure. Where nip pressure is to be produced, the oil pressure in the opposing elements is lowered. In this way, low pressures are all that is needed, and no moving elements are contained in this roll.

Through finely controlled localized adjustments, this roll can maintain a uniform pressure profile across the width of the machine to produce uniform smoothness of the sheet.

Either the covered roll or the chilled iron roll can be the swimming roll in the mat-on-line system.

The mat-on-line covered rolls are of the double wall type for temperature control. They are cooled by tap water which enters the nip at a temperature of 86 degrees fahrenheit.

The swimming rolls are externally loaded from the bottom to achieve fast separation and are equipped with oscillating doctors. Oscillating infra-red sensors scan the surface temperature of the mat-on-line cover to alert operators in case of a heat buildup.

The bone-hard cover of the mat-on-line roll is about half an inch thick.

Tandem single-nip calenders were used in the first mat-on-line installation at the Reflex paper mill of Zanders Feinpapiere at Dueren, West Germany. Here is a schematic of that installation, using rolls with a face width of 146 inches. The covered rolls are the top rolls and are 26.6 inches in diameter. The bottom rolls are swimming rolls with diameters of 17.7 inches. The nip pressure used on this application was 560 pli. The calenders were designed to operate at line speeds up to 2625 fpm and rolls were balanced for 3280 fpm.

Only one side of the sheet was treated in this application. Use of one or both nips was optional, depending on the degree of smoothness desired. Some grades were finished with single nip, some with both nips in tandem.

Here is a photograph of the same installation. (Slide #2).

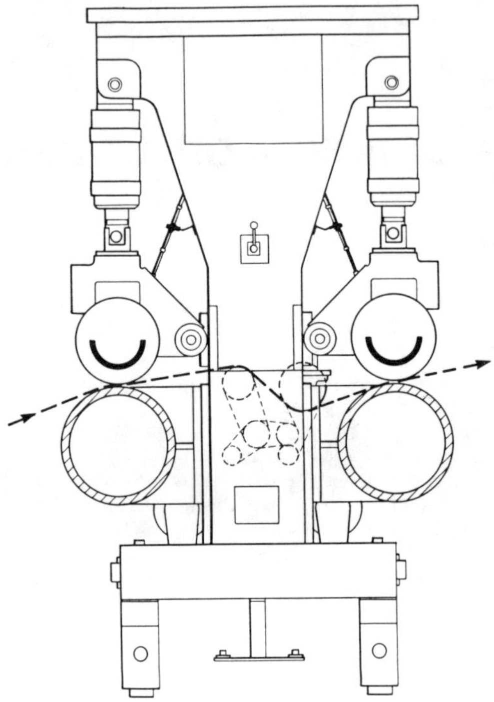

slide 2

To obtain the mat-on-line effect on both sides of the sheet in a single nip treatment, the position of the two rolls would be transposed as shown here. (Slide #3 - schematic of tandem mat-on-line calenders for treating two sides).

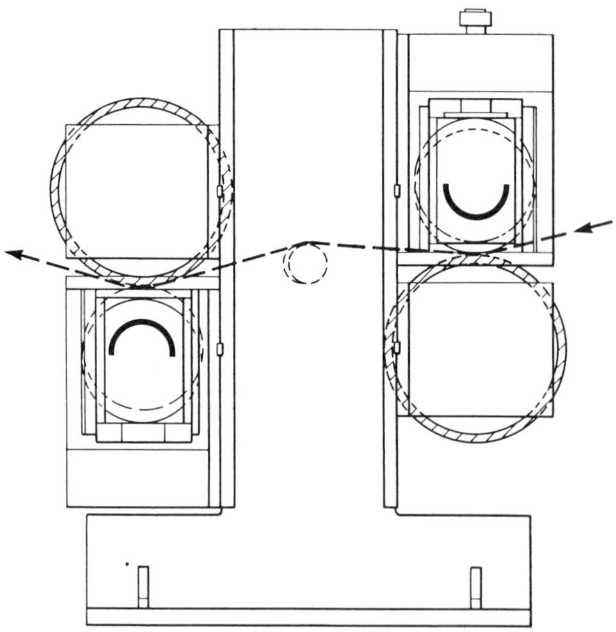

slide 3

On certain applications, it may be desirable to provide for two nips with the mat-on-line roll on each side to obtain a higher finish that the single nip provides. Tests have shown that two nips taken on the same covered roll are not satisfactory. The solution suggested by the manufacturers consists of twin tandem three-roll stacks as shown here. (Slide #4 - schematic of tandem double nip mat-on-line calenders).

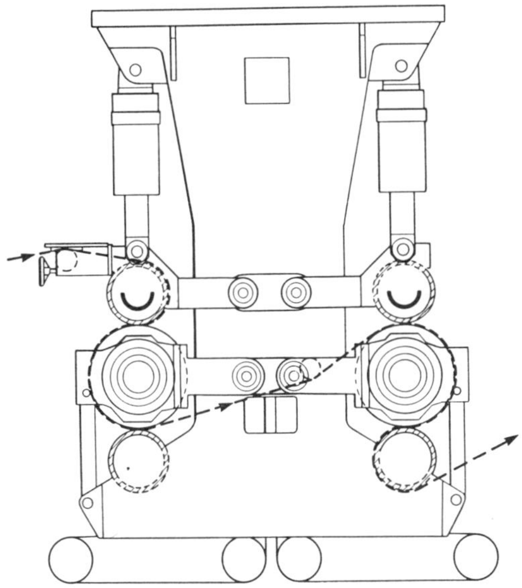

slide 4

In this case, there would be two mat-on-line rolls in each stack, for a total of four. The bottom rolls would be vario swimming rolls with mat-on-line covers. The two center rolls are shown as regular chilled iron rolls which are preferable for driving.

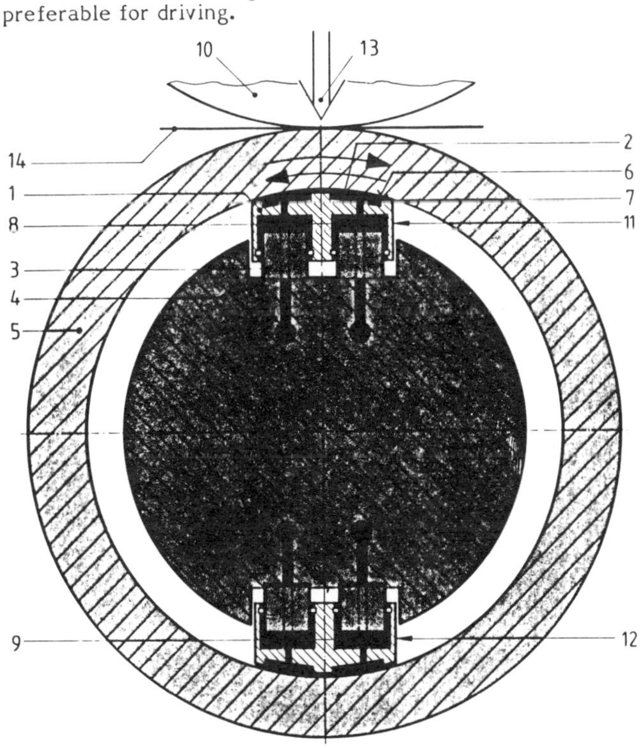

slide 5

INTRODUCE THE HYDREIN ROLL BY KLEINEWEFERS!

A report by Erhardt Muench and Wimmar Schmitz of Kleinewefers stresses the importance of the double wall chilled iron rolls with their FLEXITHERM internal heating and cooling system. Other innovations discussed in this report are: KOMBIC---a system for fast roll separation with soft landing, for use in event of a break; VARINIP---automatic adjustment of suspension spindles for quick change of calender rolls; and SENSOMAT---the center winder with controllable winding hardness, for improved roll quality and runability of larger diameter rolls.

The most important of these developments, they claim, is the heating and cooling system, since temperature "has not only been recognized for quality soft calendering and smoothness of nearly all paper grades, but has also opened up new aspects."

They found that: (1) the optimum temperature level is fixed or can be determined unequivocally, and that it can be kept constant at a selected level over the full width and length of the web and over the time span required, these operating temperatures being 20-30% higher than they used to be; and (2) the required nip pressure can now be determined more precisely with regard to optimum printability and preservation of bulk, the new findings showing that nip pressure can be reduced by up to 30%.

All the Kleinewefers machine elements and components mentioned can be integrated into new machine designs, it is claimed. All types of paper now being calendered can be treated with the new or modified calendering equipment, whether calender stacks on paper machines or off-machine coaters.

Soft calendering is seen as the answer to many of the undesirable and adverse effects of hard calendering with steel rolls only (machine calenders), such as shiny spots and specks, caused by varying bulk.

A papermakers' pipe dream until now, soft calendering could not previously be accomplished using filled rolls. But new types of fillers and roll covers has opened up alternatives in machine design and a new chance for in-line calendering, Kleinewefers says, though only when the insights gained with the "modern supercalender" are applied at the same time, namely:

1. Fast opening of roll gaps on web breaks.
2. Replacing regular chilled iron rolls with double wall designs.
3. Installing a modern heating and cooling system (Flexitherm).

The Kleinewefers literature makes this statement: "This equipment is so crucial for preserving the filled rolls that it frequently enables the replacement of the standard machine calender stacks with machine soft calenders (MSC) equipped with filled rolls."

At the same time, they are quick to point out that "there is no miracle roll which can be used universally." Suitable filled rolls must be selected by the machine builder on the basis of testing, they say. The in-line process appears to solve the quest for a suitable soft calendering roll for some grades.

Some of these designs are suitable especially for papers to be calendered with very low nip pressures, in some cases less than the dead weight of the rolls. It is the desired smoothing effect that determines the number of rolls and type of installation.

Machine soft calenders are recommended for:

(a) Papers which have been "hard" calendered up to now in machine calenders and which have displayed typical and unavoidably uneven finishes.

(b) Papers which have been "kiss" calendered in supercalenders, that is with low nip pressures and few nips (for example, hanging stock, technical papers, tracing papers, copying papers, and the like.)

The "machine soft calender" (MSC) widens the scope of concepts for soft calendering as shown here. (Slide #6 of "in-line" and "off-line" configurations.) You can identify the well-known "off-line" supercalenders on the left side (A-D), and some possible concepts of new machine soft calenders on the right side (E-H). The stack with 10-12 rolls (B) is known as the standard supercalender. (Slide #7). The 14-16 roll stack (A) is added not only for historical reasons, for example supercalendering of coated art (enamel) papers; renowned users of such calenders still keep in mind the important parameters required to achieve maximum printability of such high class papers.

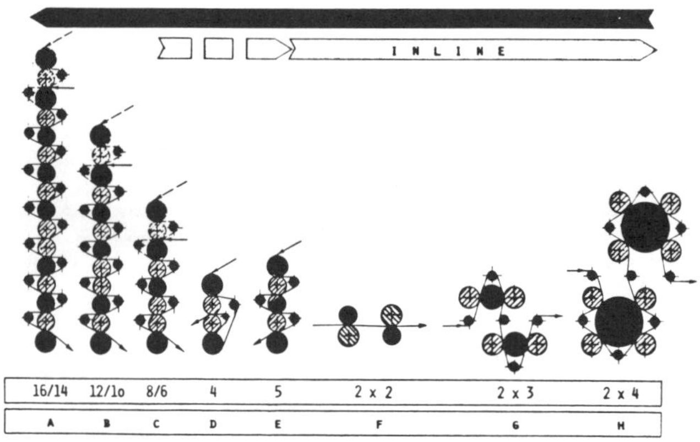

slide 6

3

slide 7

These parameters are:

- the number of rolls
- the number of nips needed to cover roll markings
- the dwell time of the sheet in the nip

This roll configuration (D) is known for the standard dull finish calender (picture)---which has featured plastic covered rolls for some time. The present state of the art in the application of plastic covered rolls could not have been achieved without the field experience gained with dull finish calendering.

The horizontal calender (Slide #8) design (F) was created for papers requiring extremely low nip pressures. (slide #9 - Vertical Design

slide 8

slide 9

Every papermaker who calenders his papers knows today where his grades fit into the range of designs. Now, however, new variants of qualities have been made possible by the improvements in the supercalender.

When Kleinewefers extended this research to certain types of newsprint, they found that there were alternate methods of obtaining optimum printability besides the machine calender. They reported the results of two test installations which yielded similar values for smoothness, roll temperature, nip pressure and number of rolls, as follows:

1. For Mill "B", which made highest smoothness their foremost requirement, the 10-roll supercalender proved the best solution in every respect. On the basis of their results, Mill B installed two 10-roll supercalenders which proved to be technologically and economically successful, exceeding their expectations.

2. Mill "A" wanted improved and more uniform finish compared to their regular machine calender and specified a target of Bekk smoothness of 180 s. The tests confirmed that the target of 180 s Bekk could be met with two tandem 2-roll or 3-roll calenders with an elastic roll in each one. In consideration of the small number of nips, it made sense to go for an "in-line" system, especially since the paper machine speed was only about 2000 fpm. It was decided that several roll combinations, variations of configurations E-H, could be employed considering the need for fast roll changes.

Wartsila-Appleton, Inc. (Appleton Machine Company) claims that in line low gloss concept is very similar to a calender concept that Appleton Machine built about 20 years ago, though not on-line. The Wartsila-Appleton system was call the "Tandem Finisher", and it was used to make a "dull" grade of paper, with gloss in the range of 25 to 35. The Tandem Finisher used a plastic shell over a swimming roll, running against a sandblasted metal roll. If you substitute a smooth roll for the sandblasted roll and put the thing on the end of a machine, you essentially have the present low gloss calender, now being sold. It's identical.

The Wartsila-Appleton concept for on-machine soft calendering is an outgrowth of technology surrounding development of the Nomex roll. Nomex is a Du Pont aramid paper which substituted cotton (or wool/cotton or asbestos/cotton) as a filler for high temperature supercalender rolls. The process was initially promoted by KMW from 1965 to 1970, the first KMW on-line supercalender using Nomex rolls having been started up in a Norway mill in 1970.

Theoretically, the material is much more resistant to marking and can withstand much higher temperatures than cotton. Where plastic shell type rolls are more analagous to rubber-covered rolls, the Nomex rolls are more like regular filled rolls. They are made the same way.

KMW put a lot of work into learning how to fill standard supercalender rolls using Nomex material. But it's mostly used in Europe in on-line supercalenders, these calenders normally consisting of five rolls, two Nomex filled rolls and three chilled iron rolls.

The KMW installations run at much higher temperatures and at pressures close to 1700 pli.

Wartsila-Appleton began to look at these types of rolls as part of their study on what type of filled rolls could be put into supercalenders. They discovered that these rolls would not provide a high gloss at conventional supercalendering pressures and temperatures, and so they were eliminated as possible substitutes for standard supercalender rolls.

Wartsila-Appleton wasn't looking at on-line calendering, but for substitutes for the cotton materials used in supercalender rolls.

The point is that Wartsila-Appleton performed numerous experiments with off-machine supercalendering in the 1950's, succeeding in being recognized by much of the printed media. But this success was short-lived, the company's efforts being frustrated by U.S. newsprint manufacturers who didn't like (Wartsila-Appleton) playing with the supercalendering of newsprint---the matter was dropped. It was at least five years later that the North American paper industry began to seriously consider supercalendering of newsprint.

Then there was a newsprint crisis in Canada, forcing publishers to import "supered" newsprint from Scandinavia, which they learned to like. Then, in a complete turn-around, they told the domestic paper industry to either make it or they would import it.

The development of on-machine supercalendering will be fueled by two streams of papermaking developments. In one stream you get a roll that is longer-lasting but can't produce high gloss. Another technology stream from Europe is the calendering of newsprint-type sheets that don't require high gloss, but demand smoothness and good ink holdout.

The newsprint area has the biggest potential market.

Now, the market is requiring the upgrading of newsprint, a requirement fueled from Europe. And the technology has been developed here for using a roll filling that permits an on-line operation to meet that future market demand.

LITERATURE CITED:

1. Thomas S. Greiner -Jagenberg Group of Companies - "Mat-On-Line", November 1982, P.I.M.A.

2. Erhardt Muench, Dipl-Ing and Wimmar Schmitz, Dipl-Ing (FH) "The Modern Supercalender -The Basis of New Concepts for Soft Calendering." Kleinewefers GmbH.

3. Les Y. Ichinose Editor P.I.M.A. "On-Machine Soft Calendering: A Good Idea Whose Time Has Come". December 1984 P.I.M.A.

Three
Supercalendering Uncoated Paper

E. Drechsel

ABSTRACT

Supercalendering wood-containing papers for rotogravure and offset printing is accomplished by controlling the sheet moisture and filler clay contents. Supercalendering conditions: temperature, steam, nip pressure, and filled roll conditions are varied to optimize the printing surface properties with minimum reduction in paper opacity. Typical operating conditions and equipment employed at Madison Paper Industries is described.

INTRODUCTION

In October of 1981 Madison Paper Industries began production of supercalendered, uncoated, high filler containing, groundwood papers designed for rotogravure and offset printing. Myllykoski Oy, a Finnish producer of supercalendered and light weight coated papers, along with the New York Times Co. sponsored the $197 million expansion at Madison, Maine. European and North American technologies and management practices were successfully married to construct and operate the only mill in North America fully dedicated to the production of uncoated supercalendered papers.

Papermaking Equipment

The major elements of the expansion consisted of a 7.75 meter (305 inch) wide Valmet paper machine, two 7.37 meter (290 inch) supercalenders, a 2440 meter/minute (8000 fpm) Wartsila winder, and the first pressurized groundwood mill in North America. Paper machine furnish is composed of 50-70% bleached spruce-fir pressurized groundwood, 18-25% purchased bleached softwood Kraft, and 12-25% filler clay. Basis weights range from 44-59 g/sq.m. (30-40 lb/3300 sq.ft.). The Fourdrinier paper machine is designed for 1060 m/min. (3500 fpm) on a 52 g/sq.m. (35 lb/3300 sq.ft.) sheet. The "Symnozzle" hydraulic headbox, equipped with an air padded equalizing chamber, is followed by a 23 meter (75 foot) long forming table. Double layer plastic wires in conjunction with a six-blade forming board, 16 adjustable single-blade foil units, four three-blade foil units, one six-blade vacuum foil box, two low vacuum wet suction boxes, and eight dry suction boxes assist in gently removing water from the furnish while retaining the filler clay. The roll-out Fourdrinier is followed by a four roll three nip press, forty-six drying cylinders, and a standard reel. The German manufactured 1.8 meter (70 inch) diameter dandy roll, located before the dry boxes, is one of the fastest operating in the world. At Madison Paper, the dandy roll and the V.I.B. sheet moisture profiling system are critical to the production of a high quality uncoated supercalendered sheet. Equipment missing from a typical newsprint paper machine is a machine calender.

Equipment missing from a typical supercalendering operation is a rereeler before the supercalenders.

Three Measurex sheet scanners interfaced with a Valmet Damatic control system monitor stock preparation and paper profiles. The paper machine operations, especially the moisture and basis weight profiles, have a dramatic influence on the quality of paper produced. The operations that influence the moisture, basis weight, and ash profiles must be rigidly controlled and supervised in order to produce paper with surface properties that can be improved by the supercalenders without sacrificing runability.

Supercalenders

Manning. The machine hall foreman at MPI is responsible for the stock preparation, paper machine, supercalender, winder, roll wrapping, and sometimes shipping departments. This overall responsibility adds a realization that departmental interactions do exist and insures that the winder, supercalender, and papermachine operators must all work together to produce the desired product.

The three man supercalender crew operates two supercalenders. The crew members judge how their machine is performing by observing the results of their work at the winder and in the quality control laboratory. Supercalender crews must be familiar with the recent changes made on the paper machine and the condition of their filled rolls. Temperature monitoring of the filled rolls with a hand held infrared radiation meter and general inspections of the temperature control and steam distribution systems are important responsibilities of all three crew members. The selection of filled rolls to be changed is initiated by the supercalender operator, approved by the foreman, and performed by the assistant operator and his helper. Recording of filled roll changes and operating conditions is performed by the operator. The changing of one filled roll will take the assistant operator and helper 15-25 minutes depending upon the roll position and operator ability. A productivity incentive that rewards all employees each month if the production goal is achieved helps create favorable working relationships.

Equipment. Two twelve roll Bruderhouse supercalenders process all the paper produced. Each supercalender contains six fiber rolls, four chilled iron temperature controlled intermediate rolls, and two variable crown rolls of the Farrel design. The hollow cast frames, rigid cross braces, and pile foundation have resulted in a vibration free operation. Fly rolls with a spiral groove surface are installed before every nip. The inside fly rolls are driven to assist open nip stack threading. Maximum loading pressure in the bottom nip of the supercalender is 350 kN/m (1960 pli). A quick separation hydraulic system, sheet cutoff device, and photocells all work together to rapidly open the nips in event of a sheet break. This system has successfully reduced filled roll marking caused by sheet breaks. The stacks are driven by two 447 kW (600 horsepower) motors to a maximum speed of 763 m/min. (2500 fpm). The unwind and windup stands can be oscillated. The windup is of the centerwind variety equipped with a rider roll and taper tension system. Quick spool change devices for ejecting an empty spool at the unwind and inserting an empty spool in the windup are used. A Measurex on line caliper scanner is located at the windup of each supercalender. Automatic suspension spindle adjustments for positioning the nips during a filled roll change have been well used. Elevators and walkways were designed with complete access to the stack in mind. Safety nip guards and emergency stop cords and pushbuttons were strategically located.

SUPERCALENDERING OPERATING CONDITIONS

Several known supercalendering variables will affect the quality and runability of the paper produced. The variables that are closely supervised at Madison Paper Industries include:

1. Moisture content of the base sheet
2. Surface temperature of the intermediate rolls
3. Nip pressure
4. Speed
5. Steam
6. Condition of filled rolls

Table 1 outlines the typical operating parameters at Madison on 59 g/sq.m. (40 lb.per 3300 sq.ft.) paper.

Table 1. SUPERCALENDERING OPERATING CONDITIONS (40 lb/3300 sq ft paper)

Parameter	Value
Base Moisture	7.5%
Steam	
Quantity	1.1 t/hr (2500 lb/hr)
Temperature	130 C (260 F)
Pressure	170 kPa (25 psig)
Roll Water Temperature	
Top Zone	100 C (210 F)
Bottom Zone	60 C (140 F)
Bottom Nip Pressure	285 kN/m (1600 pli)
Operating Speed	600 m/min (1950 fpm)
Tension	
Unwind	0.26 kN/m (1.5 pli)
Windup	0.33/0.20 kN/m (1.9/1.1 pli)

Moisture

The paper's average moisture content and moisture profile have the strongest influence on almost all quality values of Madison's supercalendered uncoated groundwood sheet. Porosity and oil absorbency values are remarkably influenced. Figures 1 and 2 show the improvement in smoothness and porosity as moisture content is increased. However, an opacity loss has been observed to occurr as moisture content is increased, Figure 3.

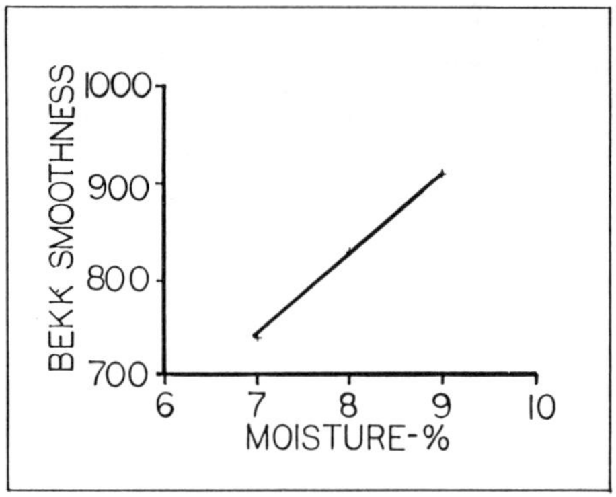

fig. 1

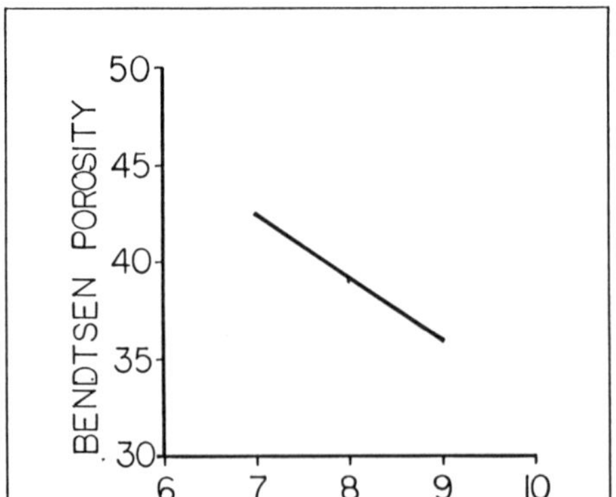

fig. 2

Temperature

Controlling the intermediate roll circulating water temperatures at 100 C (210 F), top zone, and 60 C (150 F), bottom zone improves the gloss and smoothness of the supercalendered sheet while having minimal affect on opacity. Average operating surface temperatures of the fiber and steel rolls are illustrated in Figure 4. The cool upper nip is a response to the temperature control of the variable crown roll oil. The running life of filled rolls used in positions four and five has been less than any other position in the supercalender. The high surface temperatures of the rolls are a contributing factor to their reduced operating times.

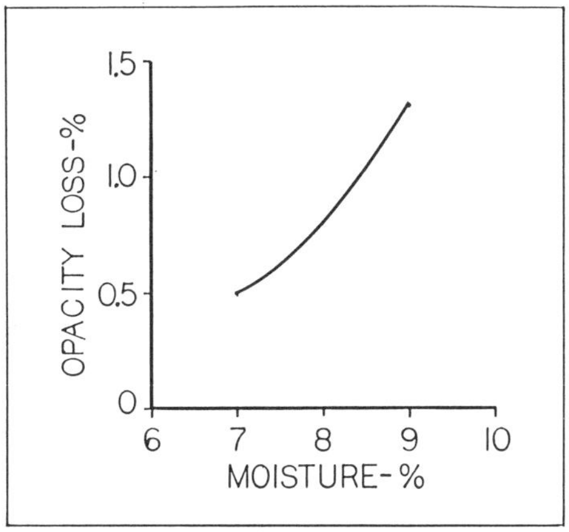

fig. 3

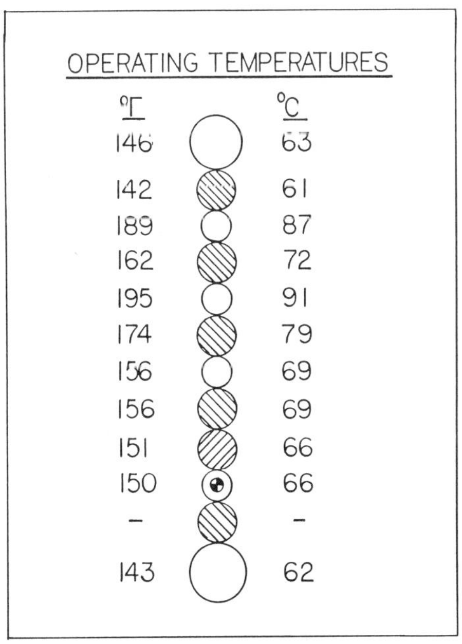

fig. 4

Pressure

Increasing nip pressures from 200 to 300 kN/m (1200 to 1700 pli) will improve smoothness and gloss, but will reduce opacity. It has been observed that as nip pressures increase beyond 300 kN/m (1700 pli), smoothness and gloss do not improve as rapidly and opacity loss increases. Basis weight and moisture profiles are critical when operating at high nip pressures. Filled roll running life will be shortened as pressures and associated temperatures are increased. Good quality supercalendered groundwood paper can be produced with bottom nip pressures between 280-300 kN/m (1600-1700 pli).

Speed

An increase in operating speed will often result in lower gloss and smoothness quality values. Figure 5 shows the relationship between smoothness and speed. The negative affect of speed increase on paper quality may be associated with the influence of steam. Increasing the amount of useable steam delivered to the paper may offset the negative influence of an increase in speed.

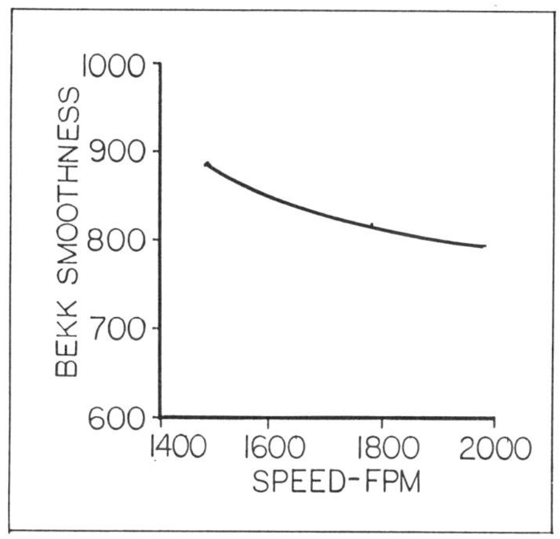

fig. 5

Steam

Steam showering that increases sheet temperature and moisture is an effective method of improving gloss and smoothness. The amount of remoistening steam to be added depends upon the desired quality, original sheet moisture, steam characteristics, and showering system efficiency. Two sidedness of a supercalendered groundwood sheet can be reduced by the proper placement and use of steam showers.

Filled Rolls

The objective of supercalendering a groundwood sheet is to improve the printing surface properties with minimum reduction in opacity. Surface properties are greatly influenced by the quality of the filled roll surfaces. In 1984, the supercalender operators at Madison Paper Industries changed nine rolls per thousand tons of paper supercalendered. This quantity of roll changes is excessive, but wet press felt hair buildup on the filled and intermediate roll surfaces required that the rolls be changed or the paper quality sacraficed. An average of six to seven roll changes per one thousand tons of paper supercalendered is typical for several European producers of supercalendered groundwood papers. Postponing or neglecting to change filled rolls will cause a reduction in paper quality and an increase in paper rejected at the winder. Monitoring the condition and temperature of the filled rolls is vital to a successful operation. Knowing the condition and profiles of the filled rolls before and after resurfacing can assist in improving a supercalender operation.

ACKNOWLEDGEMENTS

The basic experimental design and data used in this presentation for the preparation of figures 1, 2, 3, and 5 were collected and analyzed by Jukka Kejonen of Myllykoski Oy and substantiated by operations at Madison Paper Industries.

Four
Midtec's Experience with Calenders

G. Johnson

1. Midtec's Grade Structure

Midtec produces all five (5) enamel grades, as well as specialties. Thirteen percent of this production is in sheet form. Basis weights ranging from 34# to 146#/3300 s.f. are produced.

2. Review of Midtec's Production Capability

A total of six paper machines and two off-machine coaters are operating in the Kimberly Mill, in widths from 110" to 223", and in speeds from 350 fpm to 3,340 fpm.

3. Outline of Midtec's Calendering Equipment

The mill has two 2,500 fpm supers for No. 6 Complex, which calender a sheet 220" wide, two 2,500 fpm supers for No. 5 Machine (170" wide), and a single supercalender for Machines Nos. 1 thru 4 (sheet widths from 110" to 150"); three off-line, 81" wide tandem finishers provide suede and embossing finishes.

4. Midtec's Latest Supercalender Additions and Rebuilds

A 9-roll Kleinewefers supercalender, featuring Vario Swimming Rolls top and bottom, Sensomat windup, Kombic hydraulic roll separation, Flexitherm heating/cooling, and Vari-Nip suspension spindles, was installed and started up November 1, 1984. This super was part of the rebuild project, which resulted in an increase of 100 tpd and 600 fpm for No. 5 Machine. The existing 1956 vintage Beloit 11-roll supercalender was rebuilt with a new static DC drive and a multiple set unwind, and a Sensomat wind-up.

5. Operating Experience on No.4 and No.5 Enamel Grades on:

(a) a new 1984 state-of-the-art supercalender, and
(b) a rebuilt 1956 supercalender

Comparisons are made on roll and sheet quality from these supers, as well as roll change frequency, operating horsepowers, operating speeds, roll loading pressures, etc.

Five
Introduction to Escanaba's Calendering Facilities

J. Beach, D. Johnson, & E. Winchester

Part A, No. 1 Mill
Jack Beach
Converting Superintendent
Mead Corp., Publishing Paper Div.
Escanaba, Michigan

#1 Supercalender
Beloit, A Frame, right-hand, 156"
Maximum Operating Speed 1700 fpm

#2 Supercalender
Beloit, A Frame, left hand, 156"
Maximum Operating Speed 2000 fpm

These are nine-roll supercalenders each with King Roll, Queen Roll, 4 Filled Rolls, and 3 Intermediate Steel Rolls.

#3 Supercalender
Appleton, A Frame, left-hand, 168"
Maximum Operating Speed 1900 fpm

This is an eleven roll supercalender with King Roll, Queen Roll, 5 Filled Rolls, and 4 Intermediate Steel Rolls

Typical hydraulic pressures for all three supercalenders:

Low Side - 600 to 800 psi
Normal - 800 to 1000 psi
High Side - 1200 to 1400 psi

Part B, No. 3 Mill
Dick Johnson
Converting Superintendent
Mead Corp., Publishing Paper Div.
Escanaba, Michigan

#31 and #32 Supercalenders
Appleton, A Frame, 282" roll faces

Supercalenders have nine rolls each
Special features:
- Swimming Rolls, Kuesters, top and bottom
- Fly rolls driven, thread with open nips
- Four steam showers (2 front, 2 back)
- Driven at #3 position
- Unwinds and windups located front
- Elevators at front, back and each side
- Filled rolls are 32" dia. when new
- Intermediate rolls are 22" dia.
- Supers are equipped with gloss meters
- Roll change time 3-1/2 to 4 hours (full stack)

One operator, one helper, each super with a floating helper to assist both supercalenders.

Startup of these two supercalenders was October 1969. At the time of installation, these two supercalenders were the largest and most modern in the world. Top and bottom Swimming Rolls and driven fly rolls were a definite plus in the decision to purchase these supercalenders. We plan to install temperature controlled steel rolls in the near future.

Part C, No. 4 Mill
Earl Winchester
Converting Superintendent
Mead Corp., Publishing Paper Div.
Escanaba, Michigan

#41 and #42 Supercalenders
Bruderhaus, Open Frame, 298" width
Operating speed 2,500 fpm
Maximum nip pressure 2,500 pli

Special Features:
- King and Queen Rolls are Swimming Rolls
- Fly rolls are sectional, not driven
- Driven at #3 position (bottom intermed.)
- Filled rolls are 29.5" dia., when new
- Intermediate rolls are 21.65" dia.
- Gloss meters read tending and drive side and control steam showers
  Rapidtherm intermediate roll heating system
- Rapid drop system: bottom cylinders open stack on web break
- Centerwind windup with 18" rider roll
- 3 sets of photo cells detect break and activate rapid drop system
- Screwmatic system provides automatic roll separation adjustment
- Roll change time approx. 90 minutes for 4 rolls, using three millwrights

One operator and three helpers for both supers.

The open face supers were ordered to conserve space and to keep roll changing time to a minimum.

Six
Handling a Supercalender Quality Problem at the Escanaba Mill

M. Curry

The following is a record of the actions taken to improve roll quality from the two-drum winders at No.4 Mill, Escanaba:

The Problem

1. High percentage of rolls to reprocess due to offsets, slipping and run-together on the two-drum winders.

2. Excessive hardness variation across the rolls.

3. The problem was primarily on supered grades.

Description of the Problem

1. The problem could be noticed visually as gaps between the rolls of paper and the rider roll at the winder.

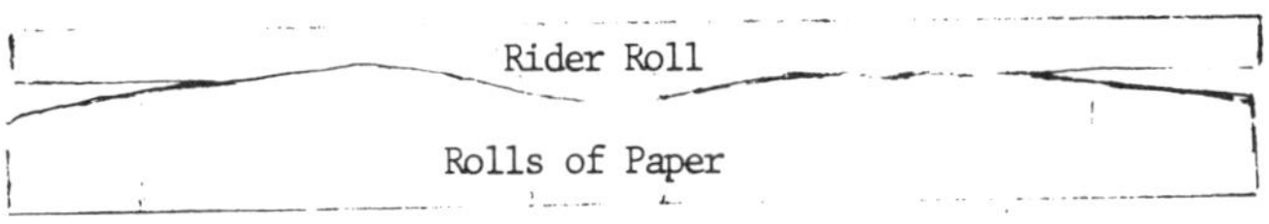

The rider roll load was supported on two hard (high caliper) areas in the paper log which was probably less than 50% of the paper width.

The side forces and tension variations across the web resulted in offsets, slips, etc.

Sequence of Actions Taken

1. Hardness profiles provided the best recordable data reference. Below is a typical Schmidt hardness profile of a problem log.

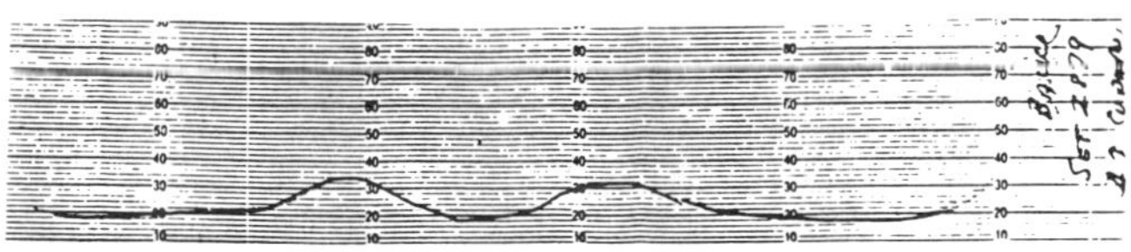

2. The hardness profile was basically the same from both 41 and 42 supercalenders.

3. Profile data were collected from the machine, coater and supercalenders:

Rawstock profiles:	Weight, moisture, caliper, smoothness, and porosity.
Coater profiles:	Coat weight, moisture, reel basis weight, and caliper.
Super profiles:	Caliper, hardness, and gloss.

Although there appeared to be initially some correlation between coater reel basis weight and rawstock moisture with the hardness profile from the supers, the correlation did not hold up as more data were collected.

Gloss profiles did follow the hardness profiles very closely. Gloss dropped in the hard areas, and increased in the soft areas. This indicated that the calender rolls were exerting less pressure in the hard areas. This was later confirmed by temperature profiles.

4. To better define whether the problem was in the supercalenders or coming to the supers, paper from #3 PM/#3 Coater was run on the #4 supers. This was done twice. The second time, freshly turned filled rolls were put into the stacks before running the paper.

In both cases, the hardness profile showed the same 'ox-bow' profile as with the #4 PM /#4 Coater.

This further confirmed that the profile was associated with the supercalender operation.

5. Nip impressions were taken on the supers to recheck and attempt to adjust crown pressures, to obtain a uniform nip at operating pressures.

6. Caliper profiles from the filled rolls from the lathe were plotted in the configuration, in which they were installed in the calender. There appeared to be some correlation between the hardness profile and the caliper profile of the filled rolls.

Also, at this time, maintenance pointed out problems in finishing these rolls versus those for 31 and 32 supers.

7. Attempts to increase crown to 'knock down' the two high areas of the 'ox-bow' profile resulted in higher losses at the winders.

8. At this point, representatives of the manufacturers of the controlled crown rolls and the supercalenders were brought in to assist with the problem.

9.1 It was found that the pressures on the swimming rolls did not come up until after the rolls were closed. This could damage the roll ends since they would contact first. The interlock was defeated to keep the swimming roll pressures on at all times.

9.2 Maintenance was asked to do everything necessary to finish the filled rolls within ± 0.001 inch. This required adjusting the lathe to cut the roll as true as possible; then, sanding to bring the roll within tolerance.

The large amount of sag in these rolls (0.125 in.) was extensively discussed. It was decided that at least 50 to 60 thousandths fill should be removed when the rolls are turned.

9.3 The temperature across the filled rolls was measures at approximately 15 inch intervals. These profiles were plotted and used to determine whether crown changes should be made in the top or bottom swimming rolls. The idea was to achieve a relatively level temperature profile (within 4°C). It was emphasized that changes should be made slowly, allowing time for temperature to stabilize before making another change. It was recommended waiting until two logs of paper were run before judging the effect of a change in crown setting.

10. The specialists remained for approximately a week to monitor the supercalender operation and continue the temperature measurements.

11. During this time, two other problems were noted:

11.1 The nip guards were contacting and damaging the filled rolls under certain conditions.

Although the conditions are not fully known, it is suspected that the aluminum guards are deflecting and hitting the rolls during the 'rapid drop' sequence.

The nip guard mounting has been modified to increase the clearance on 42 super. The change has not yet been made on 41 super.

11.2 It was also noted that the rolls do not close evenly for loading. Consequently, the rolls made contact at one end when loaded. It was felt that the weight at this time could compress the filled rolls on the end, which would show up as 'hard ends' in the paper during calendering.

This is aggravated by the Screwmatic System which requires raising and lowering the rolls several times before the system has the stop nut adjusted to the proper position.

Flow control valves were ordered, which should level the rolls for uniform contact when closed for loading.

12. Significant improvement was noted in roll quality from the winders after the work was done to: Adjust the top and bottom swimming roll crowns, improve the filled roll trueness from the lathe, and adjust the nip guard clearance on No. 42 super.

Other Comments

1. The two-drum winders are very sensitive to the hardness/caliper profile on dense super grades of paper. It is generally agreed by those in the industry that there will be a higher percentage of reprocess rolls for offsets, slips, and run-togethers on supercalendered grades than on matte type papers.

2. The best hardness profiles obtained from the super always showed some gap between the rolls of paper and the rider roll on one or both ends. However, there was definitely more rider roll contact with the paper than before (the 'ox-bow' profile).

5. Flow controls will be added in:

5.1 The load cylinders to close the rolls evenly from the bottom.

5.2 The nip load cylinders (top) to load nip pressures evenly, (currently 9 seconds difference between tending side and drive side), to time swimming roll pressures to follow nip load changes together (they presently lag by 11 seconds).

5.3 Solenoids will be added to close the control pressure valve on the swimming rolls when the rolls are not loaded (open nip). They will allow oil to flow through the swimming rolls to keep them warm. Then, as the rolls begin to close, the dead-load swimming roll pressures will be applied, which will avoid closing the stack with an undercrowned condition.

Summary of Major Points:

1. The filled rolls are difficult to finish true in the lathe. They show a large amount of sag, and it is very difficult to adjust the lathe for this.

It is anticipated that this problem will get worse as the roll diameters get smaller.

2. When turning the filled rolls, approximately 60 thousandths below the deepest area should be removed.

3. Temperature profiles should be used as a guide in adjusting crowns in the top and bottom swimming rolls.

4. Changes in stack loading should be made very slowly. The operator should wait for one or two logs (preferably two) before judging the effect of a load change.

5. Problems with the nip guards and the Screwmatic System have not yet been resolved (modification of the nip guard mountings on 42 have been completed).

Seven
SUPERCALENDERING: Analyzing Effects of Hot Calendering On The Printability Of Coated Papers

E. Munch, J. Schlunke, & W. Schmitz

The influence of temperature on smoothness and gloss-especially on the suface of coated papers-has been described a number of times in the last few years. There have also been indications of how temperatures influence or affects other parameters and paper properties.

Dr. Hunger, whose basic research in the United States was in the area of heat calenering, said at Zellcheming 1977: "The parameter of nip pressure in the paper machine calender and the supercalender has been fully exhausted, while the area of heat calendering still contains capabilities to achieve high degrees of smoothness."

The potential for better utilization of temperature in machine and supercalenders has been untapped until now because calender roll designs and conventional heating systems were unable to produce sufficiently even heating at highest temperatures and across the width of the web. Temperature was feared and was not practically utilized. Uneven calendering effects, problems with filled rolls, and difficulties with wound paper rolls-ridges, wrinkles and bursts were the result of rather arbitrary and haphazard temperature conditions.

There were instances, however, when the ability of temperature to achieve smoothness and gloss was recognized and practical applications were advanced with ingenuity and high sophistication.

The classic example is the gloss calender, whose technical concept gave rise to the greatest expectations throughout the paper industry, even for applications the inventor had never mentioned. The idea of combining a strongly heated roll with filled rolls turned out to be the critical area.

Problems which even today remain partly unsolved, as well as the experience gained during this time, were taken into account in new designs. Some of these are now being used successfully-e.g. in the heat calender; others are under development and are working in laboratory environments.

TEMPERATURE IN STANDARD SUPERCALENDERS

Steam heat: Most supercalenders still use conventional steam heat. In practical application this means it is almost impossible to calender with constant and preset temperatures. The nature of the batch operation predicates constant changes in surface temperature. As a result of changes in speed (start, acceleration, deceleration at each reel change and web break, downtime, start and renewed acceleration), the nip mechanics on the periphery of the filled roll, and thus the roll temperature, undergoes constant changes.

The uncontrolled quantity of steam supplied, or cooling water in the chilled iron rolls, heat transfer to the paper web, and heat radiated from the calender rolls precluded controlled use of temperature.

The multitude of factors influencing smoothness and gloss in the supercalender indicate very complex conditions in the nip and allow recognition of the problems which arise when temperature is to be used independently of other parameters, to be freely determined and controlled. (See Fig. 1).

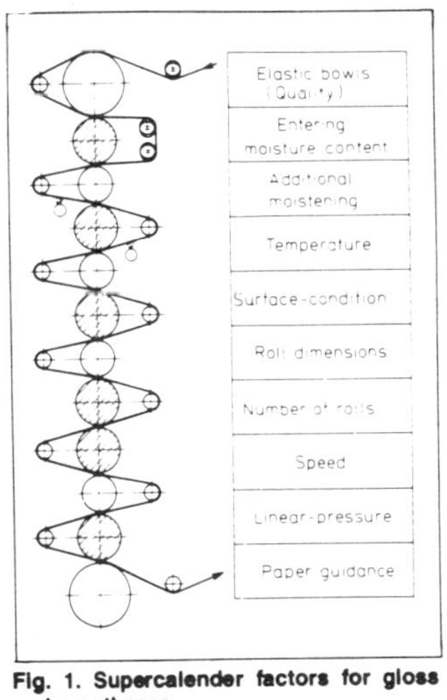

Fig. 1. Supercalender factors for gloss and smothness.

Warm water system: If the problem of scorching of the filled rolls in the supercalender is greatly diminished today, it is largely because many calenders are now equipped with warm water systems.

This system-with supplies and drains of the heating medium located at alternate sides-equalizes temperature over a time span and evens out variations over the entire width of the rolls. Operating temperatures are in the range of 40 degrees Centigrade to 60 degrees Centigrade (104 degrees Fahrenheit to 140 degrees Fahrenheit). Unwanted peaks and detrimental lows of temperature could be abated and the temperature profile across the web held constant with tolerances of about ± 1.5 degrees Centigrade - 2 degrees Centigrade (1.7 degrees Fahrenheit - 3.6 degrees Fahrenheit). This led to improved uniformity of smoothness and gloss, roll scorching was reduced, and wear of filled rolls and armatures was improved.

This warm water system serves as a heating and cooling system so that the circulating medium heats the rolls when the calender is stopped-the calender rolls rotate at thread speed-and keeps them at a preselected temperature. As soon as the calender runs at higher speed and rolls heat up because of the increased load, the result of nip pressure and speed, the circulating medium serves as cooling agent, enabling calendering in a preselected range of temperature.

Fig. 2 shows a warm water system with two control circuits, one each for the upper and lower section of rolls. This system is completely closed; the circulating water volume and thus the flow speed in the rolls are constant; only water temperature is controlled.

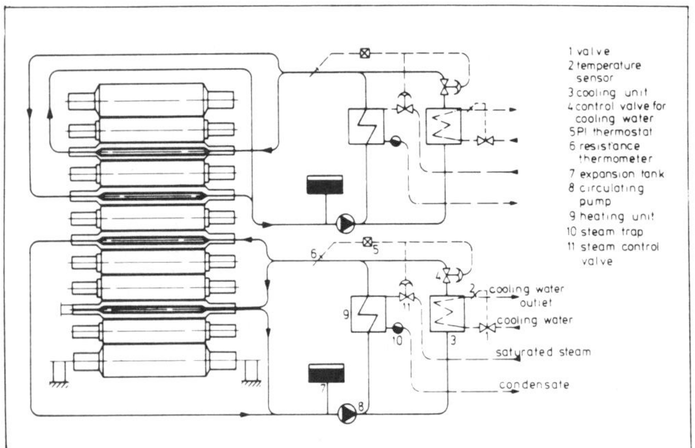

Fig. 2. Warm-water heating system (double circuit).

Fig. 3 shows schematics of chilled iron rolls. Effective roll heating in the desired form became possible only when the solid rolls with small center bores were replaced by thin shell rolls with displacement elements.

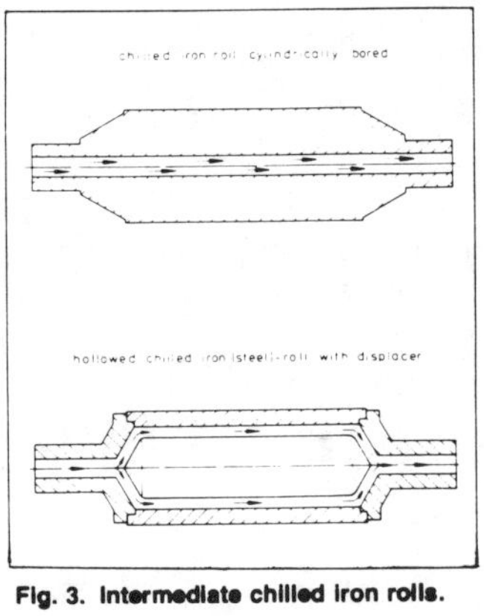

Fig. 3. Intermediate chilled iron rolls.

Heat calendering: If it was advantageous to operate supercalenders with temperatures up to 60 degrees Centigrade (140 degrees Fahrenheit), it was obviously necessary to dissociate temperature from other parameters and to focus on it by itself.

The question arose: Is high nip pressure required for its own sake to produce smoothness, or are we calendering exclusively or preferably with high nip pressures to utilize heat which evolves as a by-product of the calendering process?

An intensive test, carried out for more than a year on a modern laboratory calender at the technical center of Kleinewefers, using representative samples of comparable grades from leading European paper mills, was performed with the following idea in mind:

To make temperature control independent of nip pressure by conducting heat to the web by means of heated rolls.

DESCRIPTION AND EVALUATION OF TESTS

The following curves and tables (see Fig. 4-11) are for publication-grade coated gravure papers and show dependence of smoothness, gloss, bulk and brightness on temperature at various nip pressures. Temperatures are measured on the surface of the calender rolls.

TABLE 1
CONSTANT SMOOTHNESS WITH REDUCED NIP PRESSURE

SMOOTHNESS (BEKK)	NIP PRESSURE	TEMPERATURE
1300 sec	300 daN/cm (1680 Pli)	60 °C (140 °F)
1300 sec	200 daN/cm (1120 Pli)	80 °C (176 °F)
1300 sec	100 daN/cm (560 Pli)	100 °C (212 °F)

TABLE 2
INCREASING SMOOTHNESS WITH CONSTANT NIP PRESSURE

SMOOTHNESS (BEKK)	NIP PRESSURE	TEMPERATURE
1000 sec	200 daN/cm (1120 Pli)	60 °C (140 °F)
1400 sec	200 daN/cm (1120 Pli)	80 °C (176 °F)
1900 sec	200 daN/cm (1120 Pli)	100 °C (212 °F)

Fig. 4. Effect of temperature on smoothness.

Smoothness: The influence of temperature on smoothness achieved by calendering of all publication grade coated gravure papers-regardless of make-manifests itself in almost coinciding curves. (See Fig. 4). They demonstrate that the degree of smoothness obtained at a normal calendering temperature of 60 degrees Centigrade at high nip pressure of 300 daN/cm (1,680 pli) can be achieved at far lower nip pressures when operating at higher temperatures.

Constant smoothness: By raising temperature from 60 degrees Centigrade to 100 degrees Centigrade (212 degrees Fahrenheit), it is possible to attain the same degree of smoothness with 1/3 of the original nip pressure.

Constant nip pressure: Raising temperature at constant nip pressure can almost double the degree of smoothness at constant nip pressure. (See Fig. 4.) This relationship is approximately the same on all makes of publication grade coated gravure papers tested, while the absolute degree of smoothness may vary from make to make. (See Fig. 5.)

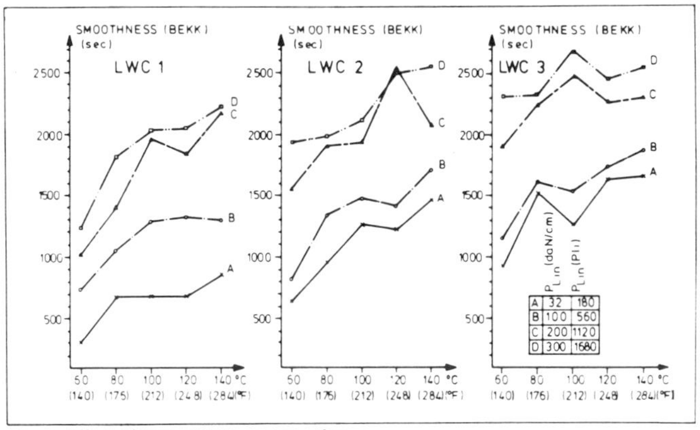

Fig. 5. Effect of temperature on smoothness.

Gloss: The curves for surface gloss are equivalent. (See Fig. 6.) The findings for smoothness can be applied here as well. By increasing temperature from 60 degrees Centigrade to 100 degrees Centigrade (See (See Fig. 7.)gloss can be almost maintained while nip pressure is reduced to one-thire of initial value (coated gravure No. 1). Stated differently; at a nip pressure of 200 daN/cm (1,120 pli), gloss can be raised from 17.5% to 26% by raising temperature from 60 degrees Centigrade to 120 degrees Centigrade (240 degrees Fahrenheit). Here again, the shapes of the curves are almost coincident, even if absolute values vary from make to make.

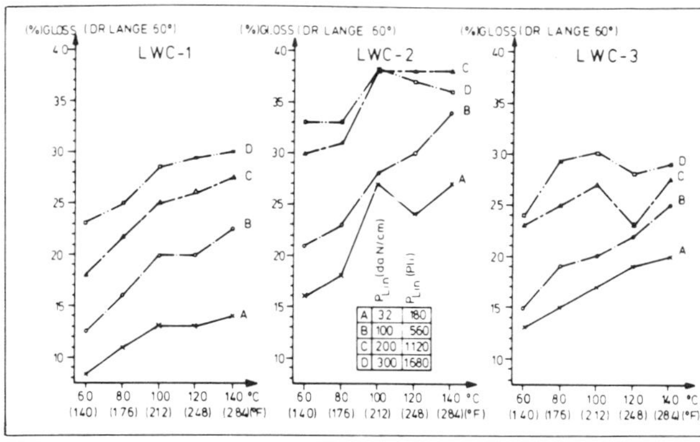

Fig. 6. Effect of temperature on gloss.

GLOSS (DR.LANGE 60°)	NIP PRESSURE	TEMPERATURE
23 %	300 daN/cm (1680 Pli)	60 °C (140 °F)
21,5 %	200 daN/cm (1120 Pli)	80 °C (176 °F)
20 %	100 daN/cm (560 Pli)	100 °C (212 °F)

Fig. 7. Effect of temperature on smoothness.

Specific bulk: It turns out that the influence of temperature on bulk is at least as strong as that of nip pressure, for which this was much more expected. (See Fig. 8.) This interconnection varies considerably with papers of different origin.

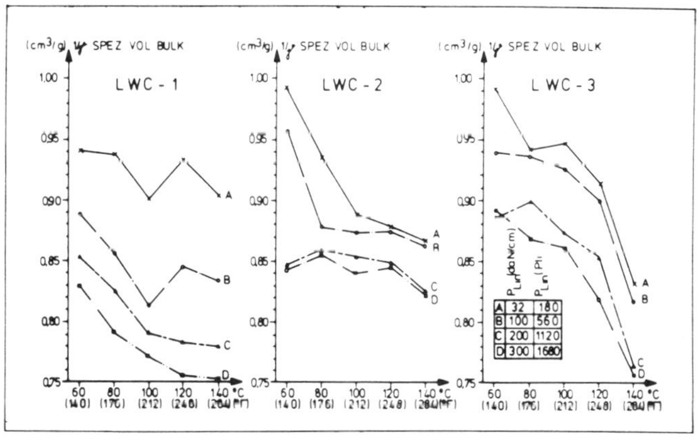

Fig. 8. Effect of temperature on bulk.

Brightness: Brightness of the calendered sheet is found to be influenced less by temperature than by nip pressure. As shown by Figs. 9 & 10, brightness of the sheet remains almost constant at rising temperatures, while there is marked separation in the curves at various nip pressures. We can conclude that brightness can be improved at higher temperatures and reduced nip pressures.

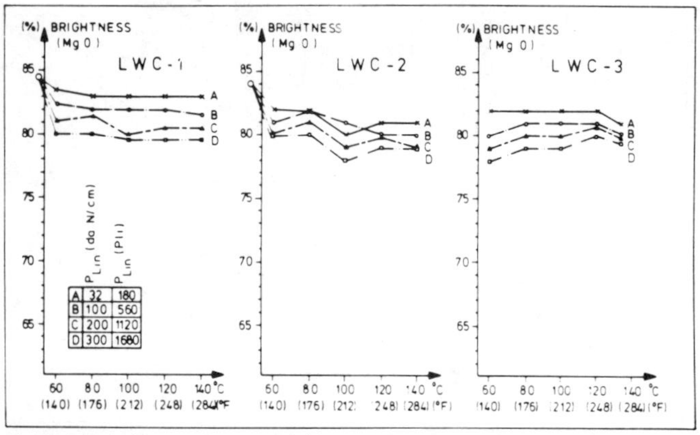

Fig. 9. Effect of temperature on brightness.

BRIGHTNESS	NIP PRESSURE	TEMPERATURE
80,0 % /MgO	300 daN/cm (1680 Pli)	60 °C (140 °F)
81,0 % /MgO	200 daN/cm (1120 Pli)	80 °C (176 °F)
82,0 % /MgO	100 daN/cm (560 Pli)	100 °C (212 °F)

Fig. 10. Effect of temperature on brightness.

Opacity: Opacity was tested according to DIN 53146 (Elrepho Instrument of Zeiss, Oberkochen). Only slight deterioration of opacity is shown (See Fig. 11.) at increasing calendering temperatures. Consolidation of the multitude of curves in a family shows that calendering temperature exerts less influence on opacity than nip pressure.

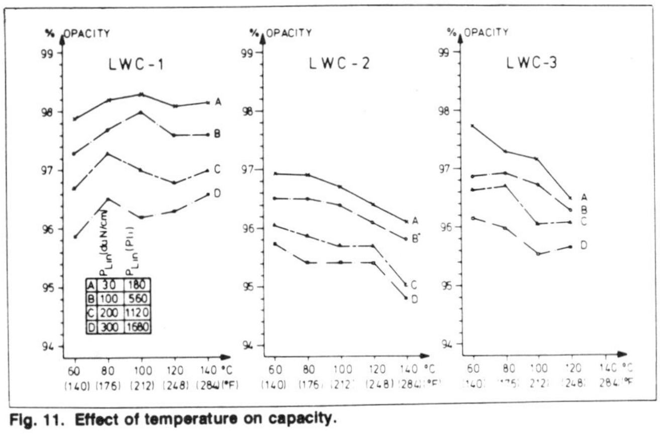

Fig. 11. Effect of temperature on capacity.

Effect on printability: Printability of the publication grade coated gravure papers was tested partly with a Helio Tester, partly with the Haindl Gravure Tester. The results (See Fig. 12) correlate to those made in test printings by the manufacturers of the samples tested.

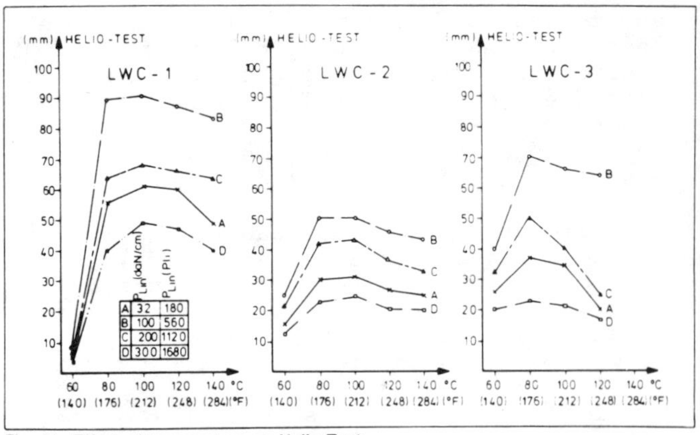

Fig. 12. Effect of temperature on Helio-Test.

INFLUENCE OF CALENDERING TEMPERATURE ON HELIO-TEST

A temperature rise from 60 degrees Centigrade to 80 degrees Centigrade (176 degrees Fahrenheit) results in a multiple of the length of the Helio test strip, showing 20 missing dots (10 mm at 60 degrees Centigrade and 300 kg/cm (1,680 pli)), compared to 100 mm at 80 degrees Centigrade and 200 kg/cm (1,120 pli)). Similar conclusions can be drawn from the tests with the Haindl Gravure Tester. The tests showed papers treated at higher temperatures but lower nip pressures have slightly rougher surfaces, but considerably improved print tests.

Out test series show that printability seems to improve at an even greater degree than smoothness and gloss. Nip pressure reduction has its limits in certain areas not coincident with minimum nip pressures of the equipment. A minimum of nip pressure- above roll dead weight, in this instance- is required to achieve optimal printability. This is demonstrated by Fig. 12, which shows the curve for nip pressure of 32 daN/cm (179 pli) to place between those of 200 (1,120 pli) and 300 daN/cm (560 pli). Excessively high nip pressure as well as excessively low nip pressure should thus be avoided to obtain optimal printability.

In connection with precise temperature control this system offers the papermaker a way of getting the optimal level of surface of his product.

Uncoated gravure paper. We conducted a similar series of tests for uncoated rotogravure papers from various manufacturers. The curves in the background show the same trends as for coated gravure papers. The rates of increase are somewhat flatter but, once again, similar for papers from different manufacturers, even though-as with coated papers- the absolute values vary from make to make.

The steps in criteria to judge printability are not as distinct as those for coated gravure papers but are still very significant. The initial values of papers from different manufacturers show the greatest differences (from]2mm to 72mm per 20 missing dots on the Helio-Test).

Some basic conclusions. Considerably improved surface characteristics for gravure printing can be obtained by calendering with higher temperatures. Reduction in nip pressure does not always result in reduced bulk loss to be expected, presumably because the heated sheet offers less resistance to a dimensional change.

Brightness of the sheet is less dependent on temperature than on nip pressure. A conclusion can be drawn that the reduction in nip pressure made possible through a rise in temperature can lead to improved brightness.

The printing tests (Helio-Test and Haindl Gravure Tester) Show the most decisive influence of calendering at high temperatures lies in printability. These improvements are far greater than those in smoothness and gloss. Thus it is possible that a sheet with hardly discernible improvements in gloss and smoothness proves far superior when heat calendered, a property not readily discernable with conventional surface test criteria. Conclusions drawn here show a trend in most papers toward increased smoothness, gloss, printability and brightness as well as loss in bulk. On the other hand, they also show that absolute values vary from sample to sample, even on comparable grades from different manufacturers.

Fig. 13 summarizes once more the values for a publication-grade coated gravure paper and shows the trends enumerated above. upon using temperatures of or above 120 C (248 F), most papers show flattening curves or even a reduction of values, especially at higher nip pressures. The reasons for this phenomenon are in the paper itself, in our opinion, and while it would be important and interesting to examine them, this area was excluded from our research.

TEMPERATURE °C (°F)	NIP PRESSURE daN/cm (Pli)	SMOOTHNESS (sec)	GLOSS (%)	BRIGHTNESS (%)	LOSS OF BULK (cm³/g)	HELIO-TEST (mm PER 20 MISSING DOTS)
60 (140)	300 (1680)	1300	23	80	-0,255	5 - 10
80 (176)	200 (1120)	1300	21,5	81,5	-0,344	102
100 (212)	100 (560)	1300	20	82	-0,356	103
120 (248)	100 (560)	1300	20'	82	-0,324	99

Fig. 13. Effect of hot calendering on paper characteristics.

From the multitude of measurements made on publication-grade coated gravure papers, it is possible to show trending curves which can be regarded as generally applicable. The display and evaluation of these curves form the basis for discussions not only among papermakers but also for the machine builder.

The results summarized in this paper show the following is possible:

(1) The hitherto compelling link between nip pressure and temperature can be varied independently of the other.

(2) Ranges of gloss and smoothness can be used to achieve optimal printability without adjusting the calender to maximum obtainable smoothness and gloss.

Both of these points carry consequences for the designer of calenders which are not limited to the design of the rolls and the a arrangement of the heating systems.

The medium range prospects are that this new know-how can be used in conventional calender design-for existing calenders as well as new machines- bu installing suitable rolls and heating systems. It would seem sensible to calender with settings circled in Figs. No. 14 & 15, i.e. with temperatures up to approximately 100 C and at nip pressures around 200 kg/cm.

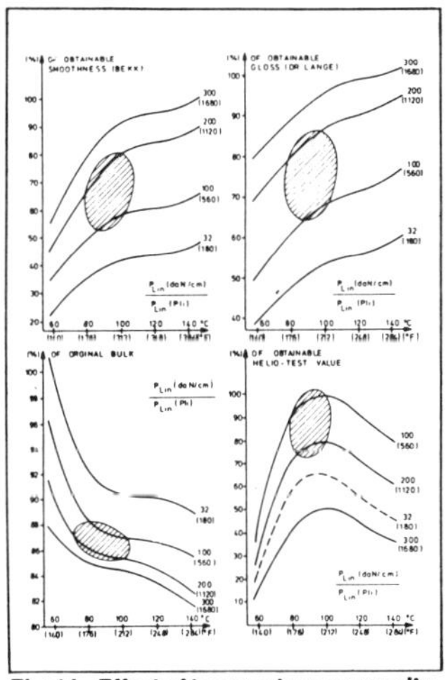

Fig. 14. Effect of temperature on opacity.

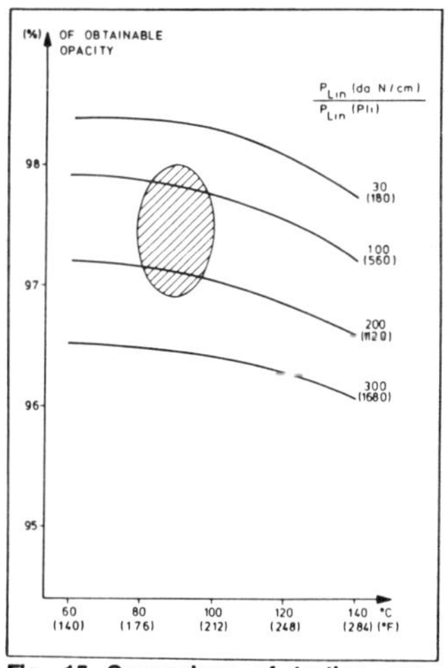

Fig. 15. Comparison of testing parameters.

Calendering at temperatures higher than 100 C does not appear to make sense for publication-grade coated gravure papers, since changes in the paper take place which should be studied jointly with the papermaker. Should it prove useful to calender at higher temperatures, such operation would have consequences for the paper as well as the calender design.

If certain papers show different curves for the Helio-Test, meaning that the optimal values of the printing tests are transposed, it may be necessary to operate with different settings. If it turns out that nip pressures around 100 kg/cm (560 pli) and less show promise, this might have more serious consequences for calender design.

For coating specialties requiring calendering at nip pressures much under 100 kg/cm and temperature ranges between 40 C and 100 C (104 F to 212 F), we have a designed a horizontal supercalender which has completed its phase of development and technical testing. Fig. 16 shows the schematic design of such a machine in the course of numerous test series.

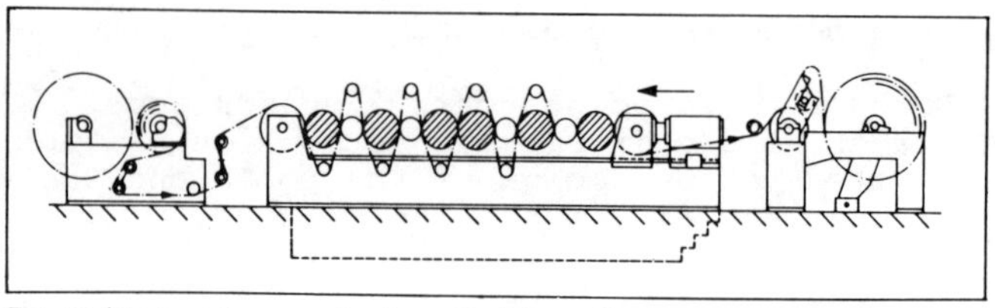

Fig. 16. Horizontal supercalender with flying splice.

Eight
Supercalendering of Coated Paper

L. Kalliola

Contents

1. INTRODUCTION

During the ongoing winter, plenty of supercalendering know-how has been exchanged between Wärtsilä and Appleton Machine Company. The information exchange mainly relates to coated grades, since uncoated grades are usually not supercalendered in North America. The running speeds and linear loadings are high seen from the European viewpoint, and therefore European filled rolls can hardly withstand the American running conditions. Therefore, a question arises, if European mills must switch over to filled rolls of the American type and to higher running speeds, as the paper-machine speeds are increasing.

This paper concentrates on comparing European and North American supercalendering modes. At the beginning, limitations of supercalendering are dealt with when it is necessary from the point of view of the comparison. At the end of the paper, information is given about synthetic filled rolls that have been developed during the last few years.

2. LIMITATIONS OF SUPERCALENDERING

On a supercalender, the quality of paper can be affected during running by means of

- linear loading
- chilled-iron roll temperatures
- paper moisture
- use of steam
- running speed

In addition, the web can be led past some nips in connection with reel changes, which takes place mainly on mat and semimat grades.

The values of the control parameters cannot be optimized merely on the basis of the paper quality values, since also the supercalendering process sets limitations. A central limitation is the heat resistance of filled rolls.

In connection with the pressing and recovering occurring in every nip, heat arises in the filled roll. This heat leaves the roll through radiation or conduction, or with the paper web to be calendered. Heat forms mostly on the roll surface, but due to the cooling occurring through the surface, the hottest place of the roll is 10-20 mm below the surface.

Increasing the linear loading and the running speed raises the nip heat, which may cause burning in the filled roll, generally at its hottest place under the surface. The highest recommendable surface temperature for wool-cotton rolls is 85-90°C. For the calender, this determines the combination of the highest possible linear loading, speed and surface temperature of the chilled-iron rolls.

The supercalendering process sets other limitations, too, for example:

Limiting factor	Limitation
Coverage of production	Minimum of running speed
Steam resistance of coating	Maximum of steam use
Maximum of end-moisture of paper	Maximum of initial moisture and steam use

3. COMPARISON BETWEEN EUROPEAN AND NORTH AMERICAN SUPERCALENDERING MODES

3.1 Calender construction

A typical European supercalender is provided with an I-frame (open frame) and 12 rolls. In North America, and A-type frame (closed frame) is more usual and the number of rolls is 9-10. A double finisher is seldom used in the U.S., although the paper would be glazed on both sides. The two-sidedness of paper is then controlled with coating, and on the supercalender with the use of steam and the winding direction. Differences between the European and North American supercalenders and supercalendering modes are shown in Table 1.

Table 1. Comparison between European and North American supercalendering modes

	Europe	North America
Calender frame	I-frame	A-frame
Number of rolls	12 (LWC) 14 (art grades)	9 10 (if with double finisher)
Double finisher	included	gen. excluded
Filled-roll composition - wool, % - cotton, %	 15-25 75-85	 - 100
Filled-roll hardness, Shore D	85-88 (LWC)	89-91 (publication grade)
Shore D	80-84 (art gr.)	88-90 (enamel)
Linear loading, kN/m	200-300	230-330
Running speed, MPM	500-700 (LWC) 300-500 (art gr.)	600-950 (publication grade) 500-650 (enamel)
Chilled-iron roll temperatures, C°	70-90	60-70

3.2 Filled rolls

European filled rolls contain 15-25% of wool and 75-85% of cotton. They are slightly softer (lower Shore D value) than the 100% cotton rolls used in North America. More nip heat arises in soft wool-containing rolls than in hard cotton rolls, and therefore, they cannot be loaded to such an extent as hard rolls. This can be seen in Table 2 that lists the results of an endurance test performed at Appleton. When the pilot calender was rotated without paper at a speed of 600 MPM and on a linear loading of 350 kN/m, burning occurred in cotton rolls after 35-52 minutes and in a wool-containing roll after 7 minutes. The corresponding destruction temperatures were 210°C and 171°C (surface temperatures) respectively.

An advantage of the wool-containing rolls is their high elasticity that prevents the occurrence of permanent marks. In the test of Table 2, 92% of the standard marks produced on the wool-roll surface have recovered, whereas the corresponding figure for the cotton roll surface is only 63%. Marking is disadvantageous especially when running coated paper, since the roll defects are easily reproduced on the coating surface. On thick and dense papers, as on art paper, the rolls are more easily marked than on light-weight and high-bulk papers.

Fig. 1 shows how different rolls affect the paper quality. When a soft wool roll is replaced with a hard cotton roll and the calendering conditions are otherwise kept constant, the paper gloss and density increase. On hard rolls, the nip is short, which means that the compression pressure (N/m^2) affecting the paper is high and the dwell time remains short. Reaching the same gloss on both rolls presupposes that the linear loading used in higher on soft wool rolls than on hard cotton rolls. The operating temperature of wool-containing rolls is then also higher than that of hard cotton rolls.

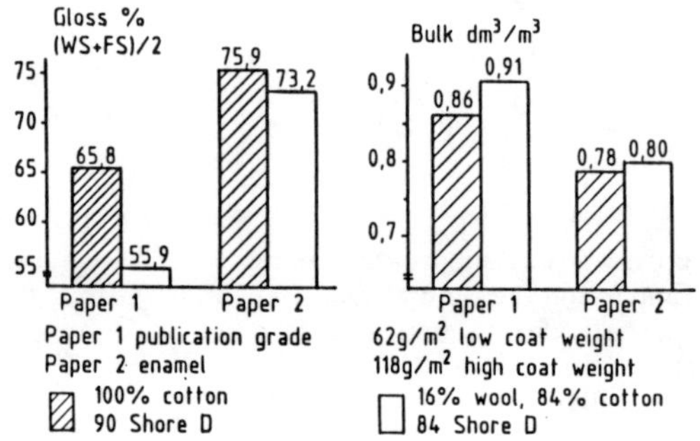

Calendering results are often measured by comparing the development of the surface properties of paper (e.g. gloss) as a function of the density. Such a comparison is well founded, since the aim of calendering is to improve the printing properties of the paper surface so that the density increase remains as low as possible. According to nip theories, a higher gloss is obtained on soft rolls than on hard ones when calendering the paper to the same density, since the shear forces prevailing in the nip are higher on soft rolls. Test results both for and against this theory have been published. Contradiction arises because the development of the gloss and the density depends also on several other factors than on the roll hardness.

3.3 Running conditions

In North America, normal linear loadings are 230-330 kN/M and running speeds 600-950 MPM, whereas in Europe the linear loadings used are 200-300 kN/m and speeds 650-800 MPM on LWC grades. In Europe, very soft rolls with a high wool content (80-84 Shore D, 20% of wool) are used for supercalendering art grades, but in production conditions these rolls do not withstand running speeds exceeding 400-500 MPM.

Table 2. Comparison between filled-roll materials

Roll material	Hardness	Mark recovery	Destruction time & temperature	
	Shore D	%	min	°C
European wool/ cotton (wool 20%)	85	92	7	171
European 100% cotton	85	76	35	216
American khaki cotton	87	78	52	221
American bleached cotton	90	63	48	210

For the time being, hot calendering is not widely practiced in North America, and the chilled-iron roll temperatures are lower than in Europe. The paper moisture content before calendering is higher in Europe, and the paper also dries more in the calendering process. The drying results from the higher temperatures and a higher number of nips used in Europe.

4. SYNTHETIC FILLED ROLLS

Several attempts have been made to replace the material of filled rolls with more durable materials, but so far, none of them has resulted in a breakthrough. At the moment, one of the most promising solutions is the Nomex roll fabricated of polyamide fibers in the same way as normal filled rolls. A surface highly resistant to marking, and a high heat resistance are combined in the Nomex roll. For example in the test series shown in Table 2, the Nomex roll did not blow out at all, but its surface temperature evened out at about 150°C during a 2-hour run. According to experiences available, the glazing effect of the Nomex roll is quite close to that of a filled roll. One of negative sides is the high price of the Nomex material.

During the last few years, Wärtsilä has manufactured a number of Nomex rolls for the paper industry. The rolls have been filled with an injection method patented by Wärtsilä. According to Wärtsilä's experiences, it is necessary to use the injection method when manufacturing Nomex rolls. Wärtsilä has, in fact, repaired some rolls filled according to conventional methods by injecting them into WEBRON rolls. The injection method was introduced at Wärtsilä's Paper Finishing Seminar in 1983.

5. DEVELOPING OF SUPERCALENDERING

In the future, developing the filled rolls further and re-optimizing the control parameters can make running speeds over 1000 MPM and a switch-over to on-machine supercalendering possible. The pilot supercalender of Wärtsilä's new research plant has a maximum running speed of 1500 MPM, and both European wool-cotton rolls and American cotton rolls are available for test runs. In addition, two Nomex rolls will be completed in the spring of 1984. The quality obtainable with different filled rolls especially at high running speeds will be studied in the near future.

LITERATURE

1. Johnson, K.: Troubleshooting Filled Rolls on Supercalenders. Paper Age 17, (1981)3.

2. Agronin, M.: Supercalender variables affecting paper quality. TAPPI 66, (1983)10.

3. Aken F., Wainwright C.: Supercalender rolls filled with Nomex aramid paper. 1979 TAPPI, Paper Finishing & Converting Conference, Nashville, Oct 22-25, 1979.

4. Paakkunainen E., Viitanen T.: Experiences of supercalender fiber rolls. 1983 TAPPI, Paper Finishing & Converting Conference, Minneapolis, Oct 16-19, 1983.

Nine
History of Supercalender Roll Filler Materials, Choices Of Materials, And Their Application

R. Bender

SESSION II - FILLED ROLLS

In this session on FILLED ROLLS we shall attempt to cover all aspects, more or less, of the subject through the use of several speakers, and a panel of - quote - experts - unquote.

We recognize that a full day could, undoubtedly, be devoted to this topic were we to go into more detail, and, if time permitted. As a result, we hope the question and answer period brings forth details not otherwise developed in the formal presentations.

The purpose of my talk this morning is to aquaint you with the selection and application of the various filler materials available for "cotton rolls", "paper rolls", "calender bowls" or "fibre bowls" - whatever you, your geographic area, or your industry calls them!

I think, therefore, that it is only fitting that we know something of the origin and evolution of the filled roll and its relationship to the development of the supercalender. You will, then, see why the data presented by my associates this morning is of such interest and importance to the finishing industry.

Since supercalendering is of primary interest to most of us in this room today, and, since we are apt to be asked to define this operation from time to time, I thought you might be interested in a description that's easy to remember and one that people outside our industry seem to readily understand.

Let's think in terms of ironing a starched shirt!
Imagine the cuff as the paper!
Imagine the starch as the coating!
The ironing board can be likened to the cotton rolls!
The hot iron can be likened to the chilled iron rolls.
The back and forth movement of the iron, and the pressure exerted on it simulates the action of the revolving rolls, under pressure, in a supercalender!
The results are the same - a smooth and glossy surface.

This is a rather elementary word picture, probably lacking in sufficient sophistication to satisfy those Engineers present today. Let's be a little more technical!

In recent years the "theory of replication" has been a much discussed theory of supercalendering action. Developed in a laboratory, the theory - simply stated - suggests that "paper surfaces replicate the surfaces of the rolls". In other words - the gloss and smoothness on the surface of any paper passed through a series of nips is a "copy, or reproduction" (according to Webster's dictionary) of the surface of the rolls. (This theory would imply, therefore, that the best surfaces on paper are actually produced by a gloss calender or a machine calender with rolls possessing superfinished, mirror-like surfaces.)

Another school of thought adheres to the definition used in Chapter 12, Volume III of "Papermaking and Paperboard Making - Second Edition" -(published jointly by CPPA and TAPPI). I quote, in part, "... attention should be drawn to the term SUPER-CALENDER. This term is used by paper manufacturers to distinguish this calender from a machine calender. In the textile industry the supercalender is called a FLAT CALENDER; - in the magnetic tape industry, a COMPLIANT CALENDER, or a SUPERFINISHER!"

"1. Calendering ... is a mechanical operation by which the web is compacted and the surface properties are improved..."

"2. Calendering action comprises three elements: pressure applied, deformation in both web and filled rolls, and temperature."

"The mechanics of machine calender nip action can be explained in the following manner: the metal rolls are made of chilled cast iron, or alloy steel. When these rolls are subjected to normal operating pressure, they retain their original contour at the nip. Under these conditions, the law of rolling friction, as applied to plastic bodies, no longer applies and within the pressure limits of operation, the nip contact area remains constant. Because there is no deformation, the property of relative motion between rolls caused by creep is absent."

The mechanics of supercalender nip action can be explained in the following manner: (Again, I quote from Chapter 12) "The behavior of filled rolls under operating conditions is a manifestation of the law governing the behavior of plastic bodies, known as the law of rolling friction, which can be explained as follows: with the application of load on the nips, the inert metal rolls cause a depression in the "plastic" rolls at the point of contact, and the "plastic" is pushed out on each side of the nip. If, now, the rolls are rotated, the "plastic material" will start to flow or creep, because of the constant effort to return to normal state. This plastic flow causes a relative motion of the filled roll surface on the metal roll surface, thus producing the polishing or friction action so essential to obtaining high finish or smoothing effect. The intensity of this action is governed by the amount of plastic flow furnished by the filled rolls and the paper passing through the nip combined with the nip contact pressure."

Now that we have given you several definitions of supercalendering action - described what supercalenders are - what they should do - and what they actually do, let's see how supercalenders evolved and examine their relationship to the evolution of filled rolls.

I am sure you are aware of the fact that this machine, called the supercalender, did not crop up overnight in some inventor's basement workshop. It, like most of our other paper making machinery, evolved from some very simple act, that at the time, was not recognized as "supercalendering," because no such term existed!

Primitive man had a very practical method of "supering" before the dawn of recorded history. Man burnished the skin side of pelts of animals with saliva, the human hand and an oval stone. The results were a surface on which to put his crude drawings. This was quite effective, serving fully to satisfy the needs of man when he had no language, and before he had other formal means of exchanging thoughts and ideas with his fellow creatures.

Man differs from all other animals on earth, in that he was born with a "built-in" hunger to communicate with his fellow man. The paper industry is truly one of the direct results of man's great drive to communicate. In order to satisfy this craving to understand, and be understood, man required material on which to record his thoughts, and, as he multiplied in number on this earth, the requirement for this material mounted.

It was natural, moreover, that man should eventually create fabrics with which to elevate himself above the animals whose skins he once wore to protect himself from the elements. It is a well known fact that the manufacture of cloth was one of man's earliest industries. After it became fashionable to dye cloth, a mechanism was required to extract excess liquid. Therefore, it is in the textile industry that we first find evidence of the "rolling calender" principle. Two wooden rollers were used to form a crude mangle, and cloth containing dye was run between them to press out the excess liquid. This was similar to a hand wringer on an old washing machine. Up through the 15th century, the operation was on a piece basis. Later - probably about 1700 A.D. - continuous dyeing came into being, and with it, the need for a more sophisticated calender.

It may appear that the origin of the "super"-calender, as such, is rather obscure, until one recognizes that it has been an evolutionary process covering a period of nearly 4500 years. When we consider that one of the most prized examples of writing in the history of man is a piece of Papyrus that dates back 2500 years before Christ, we realize that the surface on which those historic words were recorded had to be calendered in some manner - and possibly even "super" - calendered!

Paper, as we know it today, was believed for years to have been invented about 105 A.D. in China, by T'sai Lun. However, the Chinese made discoveries more recently that put the date back 150 years. Paper-like remnants were unearthed in 1973 and in 1978 in the Western Han Dynasty that conclusively establish the date at 73-49 B.C.! Paper discovered in 1974 is, up to now, recognized as the earliest relic stored in China bearing Chinese characters! The hemp fibers of this paper were actually covered with a layer of coating! This lends credibility to the belief by some historians that the Chinese first "calendered" paper in the second century A.D.; although it was not until the year 770 that Japanese artisans first conceived the idea of printing on paper.

It is known, also, that early Arabic tribes burnished paper by placing it on a walnut plank and rubbing it with a piece of agate, or flint, in order to improve the writing surface. (The improved surface was made necessary by the Arabs' finer handwriting and more viscous inks.)

However, most scholars choose to believe that Papyrus was the first of the so-called "calendered" writing materials. The manufacture of Papyrus material was a tedious operation. The reeds of the plant were cut into thin, longitudinal strips. They were then laid on a board, side by side, to the desired width, and another layer was laid on top at right angles to the first. After moistening to develop the natural gluten of the tissues, and after hammering, and drying in the sun, the sheets so formed, were smoothed or polished by rubbing with a piece of ivory, a stone, or a sea shell. So, here we see that 2500 years before Christ the principle of calendering was known, and actually, used.

In the gradual evolution from this ancient hand super-calender we find that man ultimately was forced to solve at least two major problems which, perhaps, were never fully defined. 1) A better writing material, and 2) better equipment with which to calender greater quantities of writing material.

At this point it is interesting to note that paper making did not reach Europe for more than a thousand years after the Chinese invented it. The first European mill was established in Spain sometime between 1144 and 1193 - and the art of printing did not have its inception in Europe for almost three more centuries!

In 1838 cellulose was discovered, and with this came a new and great age in the paper industry. Since cellulose makes up about 90% of such seeds as Kapok, cotton, and other flosses; 80% of such fibre as hemp, ramie, and flax; 60% of such woods as spruce and pine, and approximately 35% of straw, man had a plentiful source of material. However, he still was confronted with the problem of a mechanical means to convert this material to satisfy the ever-growing demand for writing and printing papers. He needed a means of burnishing paper on a large scale with a mechanical device, and he needed rolls that would do more than merely squeeze the paper as it passed through a nip, or nips.

The principle of calendering with so-called "resilient rolls" was already in practice in the textile industry, where rolls were made from the complete lower part of the trunk of the Sycamore tree. Hence, the term "Boles", later corrupted to "Bowls," and now known as "Filled Rolls."

Because the Sycamore boles dehydrated and splintered, man turned to metal for a more durable material. The metal, however, damaged the cloth, which, then, led to fabric (or cloth) covered bowls. This was a considerable improvement over anything developed up to that time, and Mr. David Bentley of Salford, England receives credit for being the first manufacturer of these bowls in 1784. Furthermore, he was the only one for many years!

Any obscurity that may attend the origin of supercalenders is, perhaps, due to our failure to recognize the fact that we have always had them in one form or another. But, what we did not have was a means to calender at high speeds. And, if it appears we are straying from our subject, please remember, the problem was not to find means to run rollers or bowls over paper - we had the means - what we did not have was the type of bowl that would stand up under the stress and strain of a mechanical operation, wherein heavy weight was applied through compound levers.

Man persisted, and in the 19th century developed the first bowl made of paper. It was compressed paste board. An improvement came somewhat later when linen paper was compressed to form bowls, which proved to be much better suited to supercalendering than was paste board. This development was mainly responsible for supercalender filled rolls as we know them today.

Due to the cost of linen paper, rolls made of hydraulically compressed cotton rags were next tried. However, since the surface was not very uniform, compressed rags soon gave way to carded cotton. These carded cotton rolls gave an excellent finish; but had one major fault - that was the ease with which they scorched! The first substitute material used to overcome this fault was raw wool which was treated and handled in the same way as carded cotton. These bowls were a complete failure since the wool fibers trapped air in the rolls and they became spongy and much too soft as they were used in the calender. Blending wool and cotton fibers did not cure the trouble.

In 1884 John Oliver decided that if he could overcome the problem of air pockets in raw wool rolls, wool was the material to use! After some little experimentation, he purchased the first "woollen paper" and produced rolls that were more resilient than linen paper, and, although less resilient than carded cotton, they withstood heat much better.

"Woollen paper" is really a mixture of wool and cotton fibers made into paper. This type of material for bowl making is still one of the most widely used in Europe and was used extensively in the early history of supercalendering (and textile calendering) in the United States.

Asbestos was substituted for cotton but the rolls lacked the resilience the English required for their coatings and never became popular. The Germans, however, did develop wool and asbestos content rolls. There, a strong need existed for heat resistance because they used bowls of smaller diameter that revolved faster and generated more heat internally.

German bowl makers continued to improve this type of roll, finally arriving at a furnish of about 20% asbestos fiber, 70-75% cotton and 5-10% wool. These rolls came into wide usage on glassine supercalenders where operating temperatures and pressures are generally higher than on other stacks.

(According to our records, recently "unearthed", as it were, Appleton Machine Company was filling supercalender rolls for Thilmany Pulp and Paper Company in 1913 using German-made asbestos content filler material, such as described above! and, incidentally, this is a good place to bring out the fact that Appleton Machine Company started to fill rolls for embossing and supercalendering in late 1884! It's first customers were Kimberly Clark and Fox River Paper Corporation.)

A similar mixture of blue denim cotton fibers and asbestos was used, up until 1974-1975, on rolls for glassine supercalenders in the United States. The early success of this harder, less resilient calender roll filler material (developed sometime between 1895 and 1900) led to the development of other dense (or hard) cotton fillers, both in Europe and here in the United States. (An interesting variation on "asbestos" rolls was a "laminated" roll made from about 1918 until the middle of the 70's for - principally - Thilmany, using 5 or 8 sheets of 18# blue denim to one sheet of asbestos paper ! We even tried a roll or two of 100% asbestos in the 60's.)

Cotton is, to this day, the principal cellulose fiber used in producing calender roll filler material. Raw, or carded cotton is still used in the United States for very special applications; but, most of our supercalender rolls are filled with material made from reprocessed cotton fibers. (Rolls for processing cloth and for embossing paper still use "woollen paper"!)

And, so, it seems that the development of the supercalender, through the ages, has been totally contingent upon the development of the rolls that constitute a supercalender stack. The development of the supercalender, itself, could progress no faster than man's capacity to solve the filled roll problem. This same struggle continues. Filled rolls, themselves, remain the greatest single factor controlling development of supercalenders!

What I have just pointed out is interesting - from a historical standpoint - I am sure you will agree!

However, it is a safe bet that you are not here for a history lesson; but, because today's supercalender operator is still faced with one age-old and basic problem - the marking and burning of filled rolls due to pressure and to ever increasing speed. I am afraid, it is still - unfortunately - a "necessary evil" in the operation of a supercalender! Certainly, it has been overcome to a great extent through the availability of a wider choice of filler materials in recent years; but, I think it will be a long time before the problem is eliminated!

If I were to tell you, at this point, that nothing of significance has been developed to replace cotton fibers in filled roll filler material in the past 40 years, you could immediately conclude that the industry has been derelict in its responsibilities to its customers!!! However, the fact is - nothing has been developed that will replace cotton fibers in calender roll filler material!! "What about all the excellent synthetics we have available today?!", you ask.

The fact is, synthetic fibers can be mixed with cotton fibers to a limited extent only (somewhere in the area of 2 - 4%) beyond which the quality of the filler material is adversely affected. The deficiencies of synthetic filler materials (or those containing a high percentage of synthetic fibers) can be categorized with those of plastic sleeves, cast plastic roll bodies, rubber, urethane, asbestos, fiberglas, etc. If one is developed that tolerates high pressures, it will not tolerate high temperatures. If it has a tolerance for reasonably high pressure and temperature, it lacks the qualities required for good finish in the supercalender process. (True it may smooth, but it lacks polish - if you will forgive the pun!)

Filler materials for supercalender rolls, most emphatically, 1) limit the speed of operation, 2) limit operating nip pressure, and 3) limit the ability to operate continuously, as can be accomplished with "flying splice" unwinds and "automatic" windup reels! These three factors (speed - pressure - and continuous operation) are rather obvious sources of uncontrolled heat buildup - independently! But, combined, they generate heat that is not readily dissipated by the usual components of supercalenders, and which ultimately destroys the filler material.

Therefore, it can be clearly concluded that "filled roll filler materials remain the greatest single factor controlling development of supercalenders". The development of the supercalender can progress no faster than industry's capacity to solve the problems inherent in filled rolls! In fact, you will note that the major improvements in supercalenders duscussed yesterday, and the underlying theme of today's session, not only emphasize designs to protect filled rolls during operation; but, stress the ongoing search for the ultimate filler material!

(Mr. Wallace Behnke, a Senior Research Associate with the DuPont Company for 33 years, follows me with a discussion on DuPont's contributions in the search to provide a paper-type filler material made of 100% synthetic fibers; additionally, our panel will develop dialog on experiences with other synthetic materials.

Now that we have rather firmly established the historical background of filled rolls and have shown the importance of such rolls in the development of the modern supercalender, let's see what we have to work with, today, in filler materials for rolls used in supercalendering and embossing of paper, and in finishing of magnetic tape.

Incidentally, we are indebted (for certain of the foregoing information) to Mr. R. G. Oliver, of the David Bently Bowl Works, whose great-grandfather made the first successful woollen paper bowl in England in 1884. Mr. Oliver first wrote about "calender bowls" in 1932, updated his story in 1969 - and here we are today!

* * * * *

It is not surprising that a large number of people operating supercalenders, embossers and magnetic tape calenders have little, or no concept of the construction of filled (or compliant) rolls.

Mr. Fecteau, our Filled Roll Shop Manager, will explain to you and show you the actual steps taken in the manufacture of filled rolls. Later you will tour our Roll Shop. My assignment this morning - now that the history lesson is finished - is to acquaint you with the numerous filler materials marketed today - specifically in the United States! I shall, however, touch on materials in general use elsewhere in the world. (We must recognize at the outset, however, that there are distinct differences in paper furnishes, coatings, size of orders, and markets between the United States and Canada, and the remainder of the world! It is incorrect to assume that what works well here will work equally well in Europe and vice versa!)

Approximately 25 plants throughout the world commercially manufacture supercalender, embossing calender and/or textile calender filled rolls. Some shops have limited facilities to serve a local market only. Only 5 or 6 market worldwide. At least several of these roll builders, located in the U.S. as well as in Europe, can fill rolls over 300" face width. Most of these large presses are limited to 32" diameter because of "window opening" - or design pressure. Some presses, known to the writer, have "windows" that will allow up to 60" diameter rolls to be pressed. However, there are other limiting factors that make this diameter somewhat impractical! Consequently, we visualize 36" (possibly 40") as the largest practical diameter at this time with existing presses.

Approximately fifty-five (55) distinctly different types of filler materials are used throughout the world for supercalender, embossing calender and textile calender rolls. In Europe, and those areas of the Far East that are strongly influenced by European practices, the following materials are in general use. Most come from mills in England and Germany. They range from the soft bulky long fibered raw or carded cotton - which makes the softest of supercalender rolls - through "cop cotton" (for embossing), to soft and hard grades of FilMat; cotton and wool mixes (using 5% to 60% wool); cotton and asbestos mixes (using 15% asbestos and more); through blue denim and 80# hard white papers to linen papers (100% flax) for high pressure friction calendering and calendering of cigarette papers. Probably the most widely used filler in Europe today is "wool/cotton paper". Its furnish is, typically, 13% wool, 76% cotton, 3% rayon and, surprisingly, 8% "plastic" - often unidentifiable!

We believe, only one U.S. roll builder continues to provide "raw cotton" rolls. The process is intricate and laborious, resulting in very expensive rolls. Only a few raw cotton rolls are used in the U.S. today, as reprocessed cotton materials provide comparable life and finish at much less cost. To our knowledge, no raw cotton fillers are imported; but, some blue denim, flax and woollen paper materials do come in from Europe.

Most supercalender roll filler materials used in the United States are 100% cotton and are made on conventional fourdrinier paper machines. Six (6) paper mills supply the entire U.S. market and also export some of their material to foreign roll shops. Basis weights of U.S. made "paper" fillers range from a high of 205# (17x22-500) to a low of 14# (4 mil), with 80# and 90# being more normal weights. Textile mill clippings of blue denim and white terry cloth make up the majority of the raw materials. However, AMCO's proprietary filler is made of 100% long fiber white muslin cotton. Most light blue or green fillers on the market use dyed fibers, while the blue denim fillers are mostly (but not all) actual blue denim fibers which - even today - are considered to be "unbleachable"!

These materials are normally sheeted by the paper mill, cut into octagon shaped discs and shipped to the roll builder on skids. Octagons are usually purchased 1/2" larger across the flats than the specified finished diameter of the supercalender roll.

A second type of cotton, and/or cotton and

wool filler material is used extensively in the United States, for supercalendering, embossing and, expecially, for textile calendering. Only one company manufactures (and exports) this unique material. This material, comprised of short fibered recycled cotton fibers with added cotton linters, is made on a modified Fourdrinier-type machine. Varying densities of the more common grades are obtained through compaction into "mats" of several thicknesses. The thinner the mat, the harder the material! A complete range of hardnesses is available from a soft, fluffy grade that simulates raw cotton to a board-like mat with supering characteristics similar to the thinner filler materials made on standard paper machines. Recently, materials closely corresponding to the 80# and 90# sheets mentioned above are being produced on this special machine.

These mat-like materials are ordered by the roll builder with appropriate holes and keyways cut in each "doughnut", as they are called in the trade! Outside diameter is usually 1/2" larger than anticipated finished diameter of the roll. Less in-plant preparation is required by the roll builder with this type of die-cut material.

In 1979 Appleton Machine Company introduced a proprietary line of materials for filling rolls, known as AMCO Fillers. This line was developed for the express purpose of proving that a superior filled roll can be manufactured if a rigid material specification is formulated and quality is closely controlled - from the time of selection of raw materials through the actual pressing and machining process.

At Appleton, we have long been aware that contaminants of any nature - from the size of a cotton seed hull to a sliver of beater of Jordan bar - have been potential sources of trouble in supercalender rolls. It was felt that through consistent elimination of these undesirable particles, so frequently found in materials available from existing sources, that we could produce rolls that provide extended life and increased production of supered paper per roll.

The basic difference between the AMCO material and other cotton, or cotton/wool blends is that it is cleaner and much more uniform in formation and color than other commercially available fillers. We all know that contaminants of any sort in a supercalender roll generate additional heat in the area of the contaminant; thus, producing a condition that often leads to premature failure of the roll.

Both AMCO White and AMCO Blue fillers are manufactured from 100% pure "muslin cotton". This is the cleanest, strongest, and most uniform fiber that can be used in the manufacture of filler materials for supercalender rolls. This long fiber material is hand sorted to eliminate inferior grades of cotton. It is passed through metal detectors and is, then, processed on a Fourdrinier paper machine under rigidly controlled conditions to produce the most uniform sheet of cotton paper possible.

* * * * *

At this point I would like to take a moment to place special emphasis on AMCO Blue Denim!

In the process of developing AMCO White and AMCO Blue materials, we concluded that we needed, also, a "Blue" filler that would be less expensive than our primary materials. Its place, in the scheme of things, would be in less severe applications where cost was the important consideration. It would compete directly with Fox River Blue Black (which was still being made at that time).

As we worked on perfecting a furnish that satisfied these goals, we discovered that we could readily meet our low cost criteria; but - the quality of the filler material exceeded our specifications in practically all physical tests! As a matter of fact, with very few changes in our process, we found that we now have a relatively low cost filler with service characteristics equal to our more expensive primary fillers.

The difference is, of course, in uniformity of color. (Difference in fiber length has had little bearing on physical properties). AMCO Whites are unbleached 100% "Muslin Cotton". AMCO Blues are "dyed Muslin Cotton" - dyed to a uniform blue at the paper mill.

The less expensive AMCO Blue Denim is made of 100% cotton, "unbleachable Blue Denim cuttings". The resultant color of the paper is only as uniform as the batch of cloth used to make the slurry. Consequently, rolls made of AMCO Blue Denim are not as uniform in color as the AMCO Whites and AMCO Blues. Nevertheless their ability to furnish paper and their "roll lie" factor are comparable.

* * * * *

Since we will be discussing hardness recommendations for supercalender rolls, this is a good spot to inject two additional comments that may be of interest.

Roll builders (and users alike) have long recognized the unreliable readings obtained when using so-called "hardness" testers for cotton rolls. For years the Shore Instrument Company Durometer, with "D" scale, was the only recognized measure of hardness for such rolls. (This instrument can create further confusion because it is calibrated to a 50 test block, normally while roll builders have always used an 85. There is an important difference!) In more recent years, attempts have been made to adapt the Rho-meter and the TMI Impact Tester to this application. All of these instruments are only as reliable as the individual is experienced in its use, and can provide acceptable readings only when the roll is new - or newly resurfaced with all "crust" removed! None truly and accurately identify "density" of the roll - density being the ratio of unit weight of filler material to unit volume. Many roll manufacturers, internally, are now basing their quality control and roll "hardness" on density rather than on Durometer indications. Nevertheless, mills still find it more convenient to use Shore Durometer "D" designations!

Operators of supercalenders have, for decades, used "number of hours of operation in the supercalender" as a measure of filled roll life (i.e., usable productive life between refillings). Their records were many times inadvertently, or even intentionally, designed to show only "time in supercalender", which ignores shut-downs for reel changing, sanding (heaven forbid!), filled roll changing and other sources of non-productive time.

As operating speeds increased over the years, complaints began to arise alleging loss of filler roll life, and, as more operating minutes per day continued to be squeezed out of existing equipment, the old system of recording filled roll life continued to distort facts, in many instances to an alarming extent! As supercalender designers and builders, and as filled roll manufacturers, we are attempting to impress upon users the fact that tonnage of saleable paper produced per filled roll is the only true measure of its so-called "life".

In choosing supercalender roll filler materials, the principal consideration is the type of operation. These, we can lump into four (4) general categories.- Glassine types (which include Grease-proofs and other dense papers such as Condenser and Vegetable Parchment) - Uncoated Publication Grades - Coated Book Papers - and Specialty Coated Grades, sometimes know as "Art Papers".

Let's discuss the operation requiring the hardest, or most dense filled rolls, first, This, of course, involves the supering of Glassine and other very dense, transparent, or semi-transparent grades. Glassine supercalenders normally have 13, 14, 15, or 16 rolls, (and in Europe as many as 20). These stacks operate at very high nip pressures using steam to heat the chilled iron rolls. Although design parameters usually provide for up to 300 degrees F., operating temperatures are normally maintained between 225 degrees and 250 degrees F. Operating pressures are many times as high as 3500 pli, and operating speeds to 2500 fpm are not uncommon! Densification and transparency are the objectives in Glassing supering. Blue Denim, Light Blue, Green and White - all in an 18# basis weight - are the preferred fillers in America. Roll hardness is usually 91-92 Durometer "D". Materials commonly in use are AMCO, Crane and Parsons.

Continuing with the dense filler materials, we turn to Uncoated Publication Grades. These papers are used in printing Sears', Penney's, Ward's, Spiegel's and other major mail order catalogs where 30# to 40#/3000 sq. ft. papers are specified to reduce mailing costs. Many major consumer magazines use these grades except for the cover stock.

We must, also, include Supercalendered Newsprint and Rotonews in this category.

These uncoated grades have a high percentage of groundwood fibers - either stone groundwood - refiner groundwood - or both - or they amy contain TMP or CTMP - and since a high gloss sheet with good smoothness is the typical specification, high supercalendering nip pressures and elevated temperatures are necessary to obtain required finish.

Here, too, roll filler materials in general use are Blue Denim, Light Blue, Green or Whitein 18#, 22# or 80# basis weight pressed to 90-91 Durometer. AMCO, Crane and Parsons are the most commonly used; but, some FilMat grades are gradually finding their way into this area.

Some "dull" grades, and other coated and uncoated sheets are supered with the harder, more dense rolls by the simple expedient of reducing nip pressure, numbe of nips, amount of steam, and level of temperature. While smoothness can usually still be maintained, gloss is usually reduced.

Coated Book Papers - either "on-machine" coated or "off-machine" coated - vary so greatly in characteristics and in basis weights that a much wider range of filler materials must be considered.

Here, the heavier basis weight fillers are preferred. However, the mat-like "doughnuts" made by Filter Materials, Incorporated are frequently used, also. The choice is contingent upon speed of the supercalender, width, type of coating and formulation, uniformity of caliper and moisture, and end use of product.

The 65#, 80# and 90# cottons that are marketed by AMCO, Parsons, Crane, Riegel and Simpson dominate the field. The types available range from AMCO 80 White, Blue and Blue Denim through Parsons' 65# and 90# White and Green. Crane makes 80# White, Light Blue and Blue Black, Simpson makes 80# grades, and Riegel produces 205# Extra Soft White and 85# LLT! Quite a confusing array of colors and suppliers - especially if one includes the new grades of Fil Mat!

These grades complete the category of cotton "paper" fillers. They are normally pressed in the range of 89 to 90 Durometer "D". Several of these fillers, because of their quality and uniformity can be pressed as soft as 88 "D", or as hard as 91 "D". In so doing the same filler (not the same roll necessarily) can be used to produce dull finishes as well as high gloss, enamel coated finishes.

The standard FilMat grades #5 and harder, are also used to process Coated Book Papers; but, due to their bulk and fiber length, they do not press into rolls quite as dense as the other cotton materials - being limited to 86 to 89 Durometer "D". Supercalender operating speeds and pressures are, therefore, limited somewhat in comparison! These short fiber, softer, bulkier, (or less dense) FilMat materials generate more heat at given speeds and pressures than do the loner fiber "paper type" filler materials. Operating parameters of 1500 pli and 1500 fpm are reasonable to expect with #5 and #6. #8 FilMat will operate at 1800 fpm and 1200 to 1400 pli. These conditions can be expected to provide reasonably good roll life, if rolls are not allowed to develop and sustain "hot spots" for an unduly long period of time.

New grades of FilMat are being marketed at this time that, according to Filter Materials, Incorporated, are made thinner, and denser to compete directly with 80#-90# calender "papers" in ranges of speed and pressure not considered by roll builders in the past for the more standard FilMat grades.

The last category is Specialty Coated Grades - or "Art Papers." These grades of paper are coated, usually, in "off-machine" operations where heavy weights of clay coatings, latex coatings, photo sensitive coatings, emulsions, etc., are applied by roll, blade, air knife, spray, brush, extrusion, or immersion coaters. A high gloss is mandatory, as Art Papers are used for high-grade printing, especially of half-tones, where detail and distinct shading and highlights are important. High brightness and a glossy, highly uniform printing surface are usually specified.

The softer grades of FilMat are ideally suited for these high gloss (and in many instances, high bulk) grades. XL Soft at 78 to 82 "D", as well as #1 - and sometimes #2 - are excellent choices for photographic papers and other types of sensitive coatings requiring supercalendering. #3 and #4 - pressing into denser rolls at 83 to 86 "D" - process label stock, cover stock, web coated offset and other non-sensitive coatings

The softer grades of FilMat are ideal - and, in the past were considered the "standard" - for use in the narrow, slow speed supercalenders used to process "Art Papers". In the U.S., they became the acceptable substitut material in opererations where long fiber raw cotton fillers were formerly used. For that matter, some of the softer FilMat grades are replacing raw cotton in England and in Continental Europe because of labor costs and the considerable amount of "hand labor" that must go into the preparation of raw cotton before it can be fitted to a roll shaft.

In summary - we can readily note, that despite the availability of a wide choice of filler materials, there is a definite amount of overlapping of usage; that hard and fast rules cannot apply to choice of materials and roll hardnesses; and, where, at one time, there was as much art in the building of a good supercalender roll as there was science, today, production of a quality filler and a quality roll is a science!

We see a trend toward rolls filled with more dense materials in today's high speed operations with wide supercalenders. The advantage is, of course, a sharper nip producing high psi at lower pli, with a subsequent reduction in heat generated by the action of the filler material. If one can reduce or eliminate the uncontrolled generation of heat, one can more effectively use a temperature control system for the iron rolls, and, not only improve production of gloss and smoothness; but, reduce energy losses and increase filled roll life.

With quick opening nips, and web cut-off devices, marking of filled rolls - which, in many instances dictated "soft" rolls - is no longer as serious a consideration as it once was! But still a problem!

It is interesting to contemplate the fact that we machinery manufacturers sometimes find ourselves somewhat impatient with the conservative nature of the paper industry in general, often thinking how slow it is to accept change and to try our new inventions!! Here, however, is a case where finishing departments of the paper industry are anxiously awaiting the development of promising new materials for supercalender rolls, and we manufacturers, with all our apparent technology, have yet to provide anything better than a "disc" or "doughnut" made of fibers known to mankind for centuries!!

KNOWN FILLED ROLL SHOPS THROUGHOUT THE WORLD

ENGLAND (5)

1. Richard Hough, Ltd. (Manchester)
2. David Bentley Ltd. (Bolton)
3. Allen W. Berry
4. B.Y.C. Engineers
5. Holgate, Fishwick & Leather Ltd.

CANADA (1)

1. Canadian Holyoke

U.S.A. (10)

1. Amadon Machine Company
2. AMCO
3. Beloit-Wheeler, Div. Beloit
4. Holyoke Machine Company
5. Magnat Corporation
6. O&E Machine company
7. B.F. Perkins, Div. Roehlen Ind.
8. John Verduin Mach. Corp.
9. Nicolet Paper Co.
10. Hertel Machine Company

FINLAND (2)

1. Wartsila
2. Tervakoski Paper Mill

INDIA (1)

1. Eck Haubold & Laxmi Ltd. (Bombay)

GERMANY (4)

1. Friedrich-Kalanderwalzen Aschaffenburg (Main)
2. Kleinewefers Krefeld
3. Neue Bruderhaus Reutlingen (owned by Kleinewefers)
4. Kalanderwalzenfabrik Dusseldorf Bauart Eck Hilger & Co.

FRANCE (1)

1. Mecal (owned by Kleinewefers)

ITALY (3)

1. Comecart Company
   - Timavo Paper Mill
   - Arbatax Paper Mill
2. R.I.F. Company

JAPAN (3)

1. Yuri Roll Company, Ltd.
2. Osaka Roll Company, Ltd.
3. Eiko Roll Company, ltd.

SWEDEN (1)

1. Klippans Finpappersbruk
   Lessebo Bruk (three small horizontal presses)

TOTAL = 31

* * * * *

PAPER MILLS IN U.S. FILLING ROLLS

1. Nicolet
2. Rhinelander
3. Simpson (Anderson, CA)
4. N.O.W. Paper Company
5. Midtec
6. Champion (Hamilton)?

BASIC GRADES OF CALENDER ROLL FILLERS

European-made rolls:

Wool/cotton (5%-60% wool fibers)
Cotton/asbestos (minimum 15% asbestos)
Raw cotton
Cop cotton
Blue denim
Linen papers (100% flax)
FilMat cottons (XL soft through #8)
Hard white

American-made rolls:

AMCO 80 cotton (White, Blue, Blue Denim)
AMCO 18 cotton (White, Blue)
AMCO 16 cotton (White)
Fox River cotton - WCT (4 mil white)
Simpson Lee cotton - 80# (recycled fibers)
Riegel cotton - Extra Soft White (205#)
- LLT (85#)
- LLT (18#)
Parsons cotton - white (18#, 65#, 90#)
- green (18#, 65#, 90#)
- white (14#)
Crane Company cotton - 80# (white, Khaki, light blue, blue black)
- 22# (white, Khaki, light blue)
- 32# (wool/cotton)
FilMat - XL Soft
- #1 through #9
- specials (WPH, KBN, BPH, BCF, WMT)
- wool/cottons (Khaki and blue wool - 20%-50%)
- cotton/cornhusks
- 80# and 90# "Paper"

Paper Presented:
August 23, 1983
Session II
Supercalender Seminar
Valley Inn, Neenah, WI

Ten
How Filled Rolls Are Made

R. Grant

On receipt of the supercalender rolls, they are lifted by overhead cranes with slings at the journals rather than cables. Slings help prevent damage on surfaces or on the journals. Also, note that the truck illustrated has a canvas top over the trailer. Canvas protects the filled roll during shipment, and when folded back, allows the rolls to be removed more easily with overhead cranes. In a closed truck a forklift is used which can occasionally damage the box or roll while trying to drag it.

The box is designed to help protect the metal surfaces on the journals and also the filled roll itself. The roll sits on the nut or up close to the head. This prevents any movement of the roll during shipping.

The first operation of refilling is cutting off the existing material from the roll. When the last cut is taken, the remaining fill will blow off due to the locked-in pressure remaining on these rolls.

After complete removal of the fill material, the runout of the shaft is checked with a dial indicator. These runout records are kept by the roll fillers; if a serious problem is found the customer is contacted regarding straightening the roll shaft or tooling it.

Also, the bearing surfaces are checked with a dial indicator for runout and concentricity to the shaft. The shaft dimensions are checked and recorded. The entire roll shaft will be cleaned with abrasive paper or wire brush, and prepared for roll filling.

There are two general roll locking configurations. The most common design uses a nut which screws on the threaded shaft up to the face of the head. This nut is designed to accept the locked-in force of the roll after it is refilled.

The other configuration is the wedge. These are pieces of metal that are designed to wedge between the shaft and the head in such a manner that they accept the force of the locked-in pressure of the refill operation.

At this point new "heads" or rings are machined. Since they are machined smaller during each regrind new heads must be made when the rolls are refilled. Roll manufacturers machine individual heads for each roll and machine the keyways to fit the roll shaft.

At this point, the roll fillers prepare the particular fill that the customer desires for his supercalender operation. Note that Filter Materials Inc., manufacturers of Filmat, die cuts the outside and inside diameter of the fill with the proper interference fit, and the keyways so the fill is ready to put on the roll shaft. The other fill suppliers Parsons, Crane, Reigel and Simpson supply the industry with Octagons.

After calculating the proper interference fit, the octagons are then cut to the desired inside diameter, this is done by lathe or by die cutting. If roll makers didn't cut a certain interference fit the rolls would be hollow when the material is put on the roll shaft and pressure is exherted. Too much interference fit could also cause undesireable results. As the interference fit is critical to making filled rolls, there are quality control steps to monitor this all through the process.

The octagons with the centers removed are randomized to compensate for any papermachine fiber orientation and inequities in caliper.

Keyways are put in the spun fill material. A roll may have no keys or as many as three.

Roll fillers monitor the raw fill materials constantly, such as taking humidity measurements prior to being used on the presses.

Temperature and humidity can affect the condition of fill materials. Various means are employed to maintain control of the moisture content of the paper including controlled environment and oven drying.

All the major roll builders have presses where small and medium size rolls can be refilled.

Some roll fillers have one or more presses capable of making rolls 32" in diameter, up to 350 inches in face, and have 6 million pound pressing capacity. The industry has added the large presses to meet the requirements for the larger supercalenders coming on line.

The fill material is placed on the roll shafts from the top in predetermined increments and pressed from the bottom for the calculated period of time and at a given pressure. This process is repeated until the correct amount of fill is put on the roll to provide the proper density. The amount of fill, the time pressed (pressing cycle), the pressures used and the exactness of the interference fit are all paramount in producing high quality supercalender rolls.

After the rolls are filled and pressed, they are moved to a machine shop. This machine shop is where all the finishing operations will be done prior to shipping the roll.

A rough cut is taken to remove the corners of the octagons.

It is after the rough cut that roll builders can get an accurate assessment of their performance.

Before proceeding any further, Supervisors sound the roll with a hammer. A special hammer should be kept for this to eliminate any damage to the roll from defects in the surface of the hammer. Sounding the roll with the hammer is an art. The Supervisor has to have the right touch to pick up the vibrations when he is searching for hollowness.

Proper measurements, runouts, and concentricity of bearing surface to roll shaft are checked to insure a quality finishing operation.

The roll is now given its final finishing cut, which is normally performed with a cylinrical roll finishing tool.

The roll is sanded to give it the proper surface smoothness and to remove any residual tool marks.

Roll builders use a Farrel Micrometer to check for trueness in the roll from end to end.

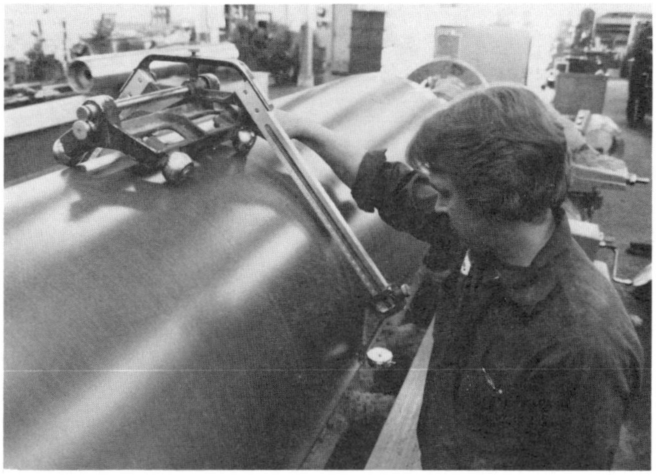

Before proceeding any further, Supervisors check the "hardness" of the roll with a hand-held Shore "D" Durometer instrument. The durometer reading should fall within an accepted range for roll hardness. Major differences in readings along the surface of the roll should be noted.

The diameter of the filled roll is checked with a micrometer or a PI tape. If a roll cleans up or finishes slightly over tolerance, it is normally shipped, allowing the mill more useable life.

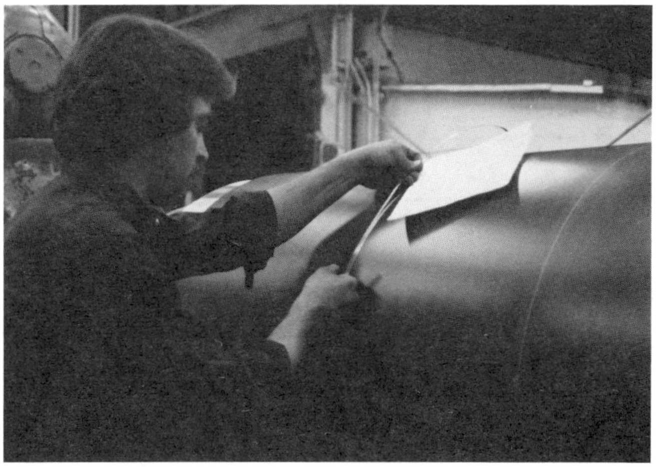

The heads and nuts are painted with a color that identifies the roll refiller. Each company uses a different color.

Now the roll is ready for shipping in the box or crate. Boxes or crates come in all shapes and conditions. Roll fillers can repair crates if needed..........

or make new wood or metal shipping boxes. Metal boxes are made in two peices for easy roll insertion. Their rugged construction insures that they will remain in service without costly repairs.

Large rolls may be shipped on cradles which cuts down dock time and reduces shipping weight.

After the finished roll is in the truck, it is secured with tie-downs and covered with a canvas to insure the quality product is delivered to the mill ready to be put in the supercalender.

Eleven
Maintenance of Filled Supercalender Rolls

R. Grant

STORING FILLED ROLLS

Storage practices for filled rolls, while they are not in the supercalender or being turned down in the lathe, vary from mill to mill. The facilities made available for storage by a particular mill often determine how rolls react over a period of inactivity.

While filled rolls are being stored, three points must be considered:

Physical Damage
Environmental Effects
Roll Sag

Physical Damage

This particular subject will be the most obvious of the three points mentioned. It must always be kept in mind that the cotton fill has the largest diameter of all components of the filled roll; therefore, a filled roll should never rest on the cotton filler. The roll should be supported by the journals, locking nuts or heads. Accessibility of rolls in storage must be considered. Overhead cranes should be all that are required for transport. Naturally, high traffic areas should be avoided to minimize damage by fork trucks and other potential hazards. Bearing lock plates should be kept on the threads when they're available. Alternatively, a heavy-duty removable tape may help prevent thread damage and burrs. Rolls are shipped from the refiller with moisture barrier wrappers to resist surface damage from oil, dirt or other possible contaminants. After inspection, many mills rewrap the filled rolls with paper which will absorb oil and water. It is recommended that the original refiller's wrapper be replaced on the rolls.

Environmental Effect

The filled rolls can go through severe environmental changes from roll filler, shipping and transfer to the ultimate storage area. The rolls should be kept in a moderate temperature and humidity environment. Excessive humidity may permeate the surface of the roll and cause softening. It has been proposed that severe moisture migrates to the bottom of a filled roll and creates unequal stresses which may result in bent shafts. High temperatures can also soften a filled roll. For example, storing a roll close to a heated pipe could result in a soft streak across the face of the roll. If this roll were put in the supercalender, the soft streak could initiate a vibration in the stack. It is not suggested that the roll be kept in a laboratory temperature or humidity environment; but the optimal storage area would be removed from some of the mill's temperature and humidity extremes.

Roll Sag

It is acknowledged that vertical storage is the best way to prevent rolls from sagging. However, few mills are equipped or inclined to store rolls in this manner because of equipment and handling requirements. The most commonly used roll turning device is a motorized stand either integral or separate from the supercalender. The rolls are rotated slowly to even out the gravitational effect on the shafts. Most mills do not yet enjoy the benefits achieved from roll turning strands, and must turn the filled rolls by hand. This should be done at least weekly, and rolls should be turned at 90°, 120° or 180° depending upon preference. Simply leave an empty space on the rack for turning all the rolls the required degree. Roll sag will be minimized by a regular program of turning rolls. If roll sag does occur, it can create problems or loss of time at the supercalender or the lathe.

When a roll has acquired a sag, it should be run out with light pressure on the supercalender. If this is not possible, the roll can be inverted or whipped in the lathe to eliminate sagging. If a roll is resurfaced with a sag, the roll will not be true when the sag is eliminated, and will most likely cause hot spots and blow-outs.

BREAK-IN PROCEDURES

An important factor toward good roll life and performance is the proper break-in (run-in) of new filled rolls. If time and conditions allow the best practice is to place a new roll in the top filled roll position of the calender stack. This position is chosen because it will have the least amount of nip loading, and stands the least change of damage. Without any pressure, with only the dead weight of the top iron (queen) roll, run the stack at speeds not to exceed 500 F.P.M. for approximately five minutes. Add around 300 pounds per linear inch pressure (P.L.I.), and 300 feet per minute (F.P.M.), at five minute intervals, until normal operating pressure and speed are reached. Practice will soon tell if this sequence can be shortened or lengthened for best results.

The type of filling is a criterion for the length of time required to break-in a roll. Softer rolls will require more attention and time during the break-in period than will the harder (90-91 Shore D) filled rolls.

Three essential functions are being performed during this break-in period:

Fill Stabilization
Surface Finish
Removal of Sag

Fill Stabilization

A newly filled roll is shipped as evenly dense and round as the roll filler can make it. Proper break-in procedures help to insure the roll is evenly densified in the supercalender before being put into service.

When a roll is refilled, the cotton discs are pressed along the direction of the shaft. While the locking nuts and heads contain the locked in pressure in the axial direction, the cotton continues to settle out in the radial direction. Breaking in a roll with moderate pressure and speed can help distribute the density and stabilize the outward thrust of the cotton discs. mills that resurface rolls in the bearing housings, prior to the break-in period, most likely eliminate or reduce this settling out. Harder fill materials seem less susceptible to diameter changes during shipping and storage periods.

Surface Finish

New rolls in particular, and even refinished rolls, require a break-in or conditioning period to develop a proper surface for finishing or calendering.

The use of small amounts of water, 50% water and 50% alcohol, or a water and pure soap (not detergent) solution, applied with a wet sponge, will expedite crusting. The roll should be sponged so that enough moisture is applied to keep the in-going nip slightly flooded. This solution should be applied evenly across the revolving roll, and as crusting forms on the surface, the roll should be run until absolutely dry.

Care should be taken to insure only moderate amounts of solution are used to accelerate crusting. Excessive moisture requires prolonged drying, can cause checking (onion-peeled surface), or can temperarily soften the roll surface.

Removal of Sag

If the roll shaft has sagged during the shipping or storage periods, the moderate run-in pressures and speeds help true the roll. Normally this sagging can be eliminate in a matter of minutes.

Never set the rolls down, that is, close the nip and apply pressure on a new roll that is not turning. It is recommended this be avoided with all rolls, but new rolls in particular could be damaged by this procedure. It will cause a flat spot that will migrate to the other filled rolls, and cause vibration in the stack.

The procedure listed above is a guide on conditioning a newly refilled roll. It is apparent that practices vary from mill to mill according to the coatings, grades, speeds, pressures and temperatures involved. Roll availability and downtime also affect the time allowed for breaking-in rolls. The moderate approach explained above could aid in increased supered paper quality and filled roll life.

CARE WHILE IN OPERATION

A filled roll will often require cleaning. Excessive water allowed to stand on a roll for any length of time can be extremely detrimental. The same solutions, used in the same manner as described under break-in procedures, are the best means of removing dust, dirt, or coating build-up after installation. Make certain rolls are run until absolutely dry.

A web of paper will compress or wear the surface of a filled roll, and leave an edge mark beyond the edges of the web which will become high. If allowed to continue, these edges will become high enough to come in contact with the metal roll. With no web in this area to help dissipate the heat, the ends of the roll will generate excessive heat and burn. A good operator will often determine how wide this deckle edge is and dub (taper) the filled roll face from the edge of the web to the end of the roll. Generally the ends of the roll will be relieved .050" and, if necessary, sanded or trimmed down again, so excessive heat will not burn the roll filling.

Sometimes a roll filler is asked to add a few inches of a more heat resistant material in this deckle area, particularly on soft roll fillings. This high edge can be a very difficult problem for mills that run different width webs on the same calender.

Scheduling wide to narrow webs is required to reduce roll changes. Other mills, running different deckle widths, will use softer rolls, which are easier to sand in the supercalender; or harder rolls, which resist the edge marks.

Another concept is softer rolls int he top, which generate more heat; therefore, pressure can be reduced, and harder rolls can be used in the bottom.

If a roll is damaged by a wrap, splice, burst or other reason, some roll fillings will recover rapidly, while in others the damage is almost permanent. If a roll cannot be removed from the calender right away, sanding or tapering the sharp edges of the damaged area may help relieve marking of the web. This damaged area will relax somewhat, and will recover to a certain degree. This recovery will occur fastest if allowed to continue to run in the stack. During severe smash-ups the fill material gets displaced and compressed, capable of causing operational problems. Displaced material shows up as high spots, adjoining the dented area, which have to be removed before further operation. Compressed material may relax during further operation, and could show up as a high spot if the surrounding area were sanded or filled.

Of course, there are many times when the damage is such that continuing to run the roll is impossible. After a damaged roll is removed, refinished and returned to operation, the damaged area could start to relax again, causing a high spot. This occurs if the cut was not below the residual compressed area of the smash-up.

Every time this high spot passes through the nit, it is being loaded beyond the normal nip load, and compressed, thus generating excessive heat. This condition is probably the largest contributor to roll burning and blow-outs. A good practice is for the roll oprator to note where this damaged area is in relation to the circumference and face of the roll, and watch for excessive heat. If the high spot is caught in time, sanding or trimming the surface even with the rest of the nip can save the inconvenience of a burned or destroyed roll. This relaxing and sanding may take a place several times before complete recovery is accomplished. The excessive heating can take place in a matter of minutes, and may continue to recur for several hours. However, here is an area in which a good operator may accomplish a lot through careful attention and good procedures.

Some operators find that a damp compress applied to the damaged area, on certain fillings, for a period of approximately 24 hours, will help relax the material. It is general practice to remove twice the depth of the damaged area from the roll. It is often better to remove enough during a refinishing operation to make certain all damaged and compressed areas are removed, than to have the damaged areas relax and cause the severe burning described above. Usually a damaged area will start to burn under the surface. by the time the area starts to discolor on the surface, it is too late to correct. Of course, if allowed to burn long enough, it will blow-out. This severe burning will then extend from at least 1/2" to more than 1-1/4" deep.

REFINISHING FILLED ROLLS

Supercalender rolls require refinishing from many reasons. A roll that has performed well in the stack and has received little damage, will ultimately require the built-up crust to be removed. Resurfacing will provide "live", resiliant cotton material fro improved supercalendering.

A wrap-up or wad from the web will normally mark the filled roll involved, sometimes seriously enough to remove from the stack. Operating loads, non-uniform nips, and edge decked marks contribute to frequent roll changes.

Roll refinishing can be segmented into four steps as described below:

Inspect Roll
Rough Cut
Finish Cut
Sanding

Inspect Rolls

Inspections and documentation of the filled roll's dimensions and any marks or damage prior to the resurfacing are valuable for operational and maintenance reference.

With the roll in this lathe, check the diameter of the cotton at the ends, quarter points and center. Smaller diameter at the ends could indicate operating pressures in excess of the crown. Conversely, larger ends could indicate underloaded crowns.

A rolling caliper will provide an approximate indication of roll trueness or straightness from end-to-end. If the roll is turned in the bearing housings, a dial indicator reading on the fill surface will provide roll sag and concentricity information.

A roll turned on journal centers can be dial indicated for trueness of roll centers as compared to the bearing surfaces.

Severe smash-up marks, burning or any other indications of cotton fill fatigue should be noted.

Chapter 8 deals with supercalender filled roll records and record-keeping of supercalender operation and resurfacing. Inspection steps will be recommended in detail, as well as the inter-relationship of the record with roll performance in the stack.

Rough Cut

A carbide tool in the general shape of a thumbnail with a good radius is probably the most used for rough cuts. This configuration minimizes tool marks and allows the deepest cut with the least chatter.

The rough cut is normally made at around 100 RPM, faster when in bearing housings, with a cutting depth of .050" to .100" per side. The lathe feed (lead) ranges .024 to .042 per revolution at most mills. Turning in bearings and housings generally allows a faster feed with less vibration than turning from journal centers.

When removing a crust build-up, the depth of the cut can be determined by watching the material being removed from harder, 88-91 Shore D filled rolls. "Live" material will tend to be "curly" or "ribbon-like" when cut, while crusty or burnt fill material will tend to be "chunky" or "powdery". The softer fill materials will not curl as they are cut, but a change in physical appearance will be evident between the crusted and "live" fill.

If a severe smash-up is evident from deep marks in the cotton fill, care should be taken to remove the compressed material under the smashed area. A common rule-of-thumb is 2-1/2 times the depth of the depressed area. For example, if a localized cut is taken at the damaged area to a depth of .050" to clean up, the roll should be cut an additional .075". After the depth is determined by "spotting", the roll should be cut to finish .125" smaller than when received at the lathe. The .125" removed should include both rough and finish cuts.

Burnt areas should be cut to the "curly" material mentioned above. Some residual discoloration may remain which does not have to be removed, but its location should be noted in the records.

Finish Cut

A shaped carbide tool or a circular roll finishing tool are most often used for the finish cut on a filled roll. RPM of 100 or more, with .016 feed are used with the carbide tool for a cut of approximately .030. Circular roll finishing tools allow increased feed and cutting depths.

The surface of a finished roll should measure within .002" Total Indicated Runout (TIR), and .002" end-to-end profile. Some mills demand closer tolerances, but roll makes generally subscribe to the above.

Sanding

Filled rolls are sanded to remove residual tooling marks. Most mills sand with a 60 to 80 grit abrasive, then finish with a 120 to 150 grit cloth-back abrasive. Various methods of sanding are used. Some examples are hand-held blocks, fabricated pads attached to the carriage of the lathe, driven or stationary belts, or sheets traversed by hand across the roll face. All can obtain a proper surface for supering paper.

After sanding, the roll's finished dimensions should be recorded and the roll surface wiped clean with a dry cloth. It is now ready for covering with an oil/moisture-barrier wrapper for protection during storage.

Twelve
A Study Course on Supercalender Filled Rolls

G. Fecteau, & R. Bender

INTRODUCTION

The following questions have been raised frequently during our discussions with our supercalender and filled roll customers.

The answers are based upon our many years of experience as filled roll builders and upon our contacts with supercalender superintendents and operators throughout the country.

Time and space do not permit us to cover every minute detail in our answers, nor to provide more than basic knowledge in some instances.

A. OPERATIONAL PROBLEMS

1. WHAT CAUSES BURNING?

Relatively uniform burning of the entire roll face is caused by heat generated in the nips of the supercalender due to pressure and/or speed. Serious, localized burning in, however, can usually be attributed to "wrap-ups", or to "wrinkles" and "tails" that have passed through the nip, causing concentrated pressure at these points. This type of burning does not generally show up until some time after the "accident" has taken place.

2. WHAT CAUSES BLOWOUTS

Blowouts are due to "wrap-ups" and "wrinkles" passing through the nip and damaging the sub-surface of the filled roll. Eventually, if the roll(s) is not taken out of operation in time, the fibers in the individual sheets of filler material break down, causing the surface to separate from the body.

3. WHAT CAUSES "FLAKING"?

"Flaking" is generally caused by an excessive buildup of crust on the surface of a filled roll. a supercalender filled roll should be removed periodically for resurfacing (even though not seriously marked) to get down to new material again. If a roll does begin to "flake", the only method of correction is to pull the roll from the stack and resurface it.

4. WHAT CAUSES "CHECKING"?

"Checking" is caused by use of excessive amount of water during the run-in process. The surface of the filled roll absorbs this excess water, trapping it below the surface temporarily, ultimately producing minor "pop out" holes.

5. WHAT CAUSES "HOT SPOTS"?

"Hot spots" results from innumerable causes. Since most of these hot areas are higher than the surrounding roll surface, pressure is localized, generating additional temperature and further increasing the expansion of the cotton fibers. These high areas are the result, in most instances, of damage to the roll surface. When a roll is "smashed-up", the filler material rises around any depression and two (2) types of damage result; rupture of the fibers in the depression area and generation of heat in the high area. Oils spots, or localized application of any fluid that produces swelling of the fibers can also produce "hot spots".

6. HOW DO YOU CHECK FOR "HOT SPOTS"?

A filled roll should be checked periodically with a Pyrometer. Either a hand-held Pyrometer, or an Optical Pyrometer can be used.

7. WHICH PYROMETER IS BEST; SURFACE OR OPTICAL?

It is more desirable to use the Optical Pyrometer as the operator does not have to make actual physical contact with the filled roll to find these "hot spots". Furthermore, the Optical Pyrometer reacts in 3/10 of a second, whereas, the Surface Pyrometer is much slower, and can be affected by friction of the revolving roll.

8. HOW OFTEN SHOULD THE SURFACE TEMPERATURE OF FILLED ROLLS BE CHECKED?

It is recommended that the temperature be checked at least once every four (4) hours of actual operation.

9. WHAT CAN BE DONE TO CORRECT "HOT SPOTS"?

The best method of correction is to pull the roll and have it completely resurfaced in a lathe or a grinder.

NOTE: Naturally, the best course of action is to do all possible to avoid "hot spots"! Equipment is available for supercalenders whose function it is to minimize (if not eliminate) severe roll marking. Systems for monitoring web profiles at the paper machine and at the coater contribute to uniformity of web to the supercalender. These modern devices do assist in improving filled roll life.

10. WHAT IS THE NORMAL LIFE OF A FILLED ROLL?

Filled rolls have been known to last as long as 38,000 to 40,000 actual running hours under what can be considered "normal" operation. A more "normal" life, however, is less than half this period; probably more like 15,000 to 18,000 hours for rolls supering publication grades and about 4,000 for glassine grades. This, of course, depends upon the filler material, upon the application, and upon pressures and temperatures to which the roll is subjected.

11. HOW DOES ONE KNOW WHEN A ROLL SHOULD BE REFILLED?

It is determined by a number of factors:

(a) A filled roll is a large, resilient mass and eventually the filler material "devitalizes". It is recommended that the roll be refilled at such time as it no longer reacts properly in the nip. "Devitalized" material can be recognized by the granulated form in which it comes off of the tool when the roll is resurfaced. If you cannot get down to a steady, curly chip, the roll should be refilled.

(b) When the roll is "worn out". This is the point at which the diameter of the roll is turned down in minimum thickness of filler material, or down to locking nut diameter.

(c) When the diameter of the roll is such that it no longer produces the desired finish or gloss.

(d) When the roll has a number of serious blowouts, or seriously burned areas, whether or not they affect the finish or the paper.

12. HOW SMALL IN DIAMETER CAN A FILLED ROLL BE TURNED?

The filled roll can safely be turned down to the same diameter as the actual locking nuts. The lathe, or grinder operator should use caution, however, so that he does not actually turn metal off of the nuts, as this weakens the locking device.

When it comes to wedge-type locking or taper nut locking devices, the roll builder should be requested to advise for each roll. This is due to lack of uniformity of design, resulting in variations in taper that cannot be observed by normal visual inspection.

13. WHAT HAPPENS IN THE "NIP"?

Since the filled roll is a resilient mass, the chilled iron sinks into the surface of the filled roll, causing deformation of the filler material with a reluctant "creep". The deceleration, followed by acceleration of the deformed surface produces a "shear" action on the web in the nip that finishes the paper. The action of the filler material is very similar to that of a car tire going through mud. It has been noted that at a given speed, with a vehicle of given weight, in mud of a given consistency, the mud will pile up in equal amounts in the front of the tire and behind the tire. This is due to the "creep" or "plastic flow" of the mud under the tire from front to back. A similar "creep" is evident in filled rolls. The hardness, or density of the filled roll determines the degree of deformation in the "nip" and of subsequent "creep".

14. WHICH SIDE FINISHES THE PAPER; FILLED ROLL OR CHILLED IRON ROLL?

There is much controversy throughout the world concerning this question. For many years, the experts were adamant in their belief that the chilled iron rolls "finished" the paper. Moving the "double finisher" nip up, or down, one or more positions in a specific stack appears to confirm this belief.

Today, it is true, many mills are finishing both sides of a web uniformly without a "double finisher" nip; but, on the other hand, many mills must still use ten (10) rolls for 2-sided coated paper.

We think the blade coater, latex coatings, and possible wires and felts of synthetic material on the paper machine contribute to the ability of some mills to do 2-sided supering in a nine (9) roll stack with one side of the paper against the filled rolls and the other against the chilled iron rolls.

To determine if your coated stock can be finished with one side running against the filled rolls and one side running against the chilled iron rolls, it will be necessary to remove the "reverse" or "double finisher" nip, add single roll "spacers" under the pressure cylinder rams, and supercalender several reels of paper, or conduct trials on a lab calender.

15. WHAT IS THE MAXIMUM DESIRABLE TEMPERATURE FOR FILLED ROLLS?

Since most filler materials for supercalender rolls are primarily cotton, and since cotton begins to deteriorate at 225°F, we generally recommend that these rolls not be run above this temperature. On the other hand, prolonged running even at 180°F will cause discoloration (actually, burning) of the surface.

140°F - 170°F is a desirable temperature range for maximum cotton filled roll life. Rolls are being operated at temperatures considerably higher than 225°F in glassine supercalenders; but, roll life is proportionately reduced.

16. WHAT CAUSES LONGITUDINAL CRACKS?

Longitudinal cracks are key cracks beginning at the shaft and working outward to the surface of the filled roll. Key cracks usually develop in filled rolls due to "old age". The material loses its resilience, cannot tolerate nip pressure concentrated over the key, and eventually a key crack shows up. The shape of the key, number of keys, accuracy of fit of keyway to key, wall thickness of filler material, and operating nip pressure are all contributing factors.

If a longitudinal key crack develops, it is recommended that the roll be returned for refilling immediately.

17. WHAT CAUSES RADIAL CRACKS?

A radial crack is the result of the separation of the individual layers (sheets) of filler material. If the filled roll absorbs excess moisture, the fibers swell, increasing the density of the roll, compressing individual fiber bundles, and possible re-orienting a portion of the fibers in the individual sheets of filler material. Upon drying again, the fibers do not return to their original form, and overall density is reduced, permitting radial cracks to form and the sheets of cotton to separate.

18. WHAT METHODS ARE EMPLOYED TO TAKE "NIP PROFILES"?

A static "nip profile" can be taken with carbon paper against another sheet of paper; with NCR-type paper or with aluminum foil.

A moving nip reading can be taken using self-contained carbon paper. It is suggested that a piece of carbon paper, whose length equals the circumference of the filled roll, be passed through the nip.

19. WHAT TROUBLE CAN BE DETECTED WITH A "NIP PROFILE"?

A static nip profile shows nip <u>width</u> and <u>uniformity</u> of nip width.

A moving nip profile provides a visual chart of the high and low spots around the circumference of a filled roll.

B. RESURFACING

1. WHAT IS THE BEST RESURFACING PROCEDURE?

Filled rolls are being resurfaced successfully in both roll grinders and lathes. We prefer to lathe-turn filled rolls using the Appleton "Roll Turning Tool", as this tool actually <u>shears</u> the fibers rather than pulling them, as is <u>done</u> in a grinding operation. The differences in finish can be readily demonstrated by running one's hand over a recently <u>ground</u> roll in a circumferential direction and then sliding it back in the opposite direction. In one direction, it will be completely smooth, while in the opposite direction, one note the fibers will stick up on the roll; not so when <u>shear</u> <u>cut</u>!

2. HOW DO YOU KNOW WHEN YOU HAVE GONE DEEP ENOUGH IN RESURFACING?

This can be determined by the condition of the chip coming off the tool. If it is granulated and broken up, you are not below the crust. If it is coming off in a steady, "curly" form, this indicates that you are below the crust. this holds true for AMCO, Crane, and Parsons grades. FilMat rolls, on the other hand, <u>never do</u> produce a steady, curly chip. The only method to use with FilMat rolls is the visual appearance of the roll surface.

Rolls must, also, be resurfaced for reasons other than mere build-up of crust. Surface marking must be removed, and only the mill can determine when its severity is detrimental to quality. The deep marks from "smash-ups" should be removed as-soon-as- possible by resurfacing. In such case, a good "rule of thumb" is to turn off an additional .125" of material, per side, <u>after</u> the deepest mark disappears. In other words, go .125" <u>below</u> the deepest mark!

3. HOW MUCH SHOULD BE REMOVED WHEN RESURFACING?

It is recommended that at least .045" per side be removed when resurfacing. This is to insure that one gets below the crust on the surface of the roll. There <u>are</u> times, however, when greater depth is required to get below this crust.

4. HOW OFTEN SHOULD A ROLL BE RESURFACED?

Whenever the crust on the roll builds up, or the roll is market severely in a number of places, causing an adverse affect on the finish of the sheet.

5. WHAT EFFECT DOES RESURFACING HAVE ON ROLL LIFE?

In many cases, periodic resurfacing actually prolongs roll life due to the removal of the crust on the surface. Prolonged operation without resurfacing creates a build-up of very thick crust on the surface of the filled roll and, sooner or later, results in removal of just as much (or possibly even more) stock during the resurfacing process.

6. WHAT TOLERANCES SHOULD :BE HELD DURING RESURFACING?

Rolls should be resurfaced within .002" from end-to-end. Many mills, however, resurface their own rolls to within .0005", using roll grinders, or lathes and lots of time.

C. MANUFACTURE OF FILLED ROLLS

1. HOW MANY DIFFERENT FILLER MATERIALS ARE THERE?

There are approximately fifty-five (55) standard filler materials on the market today. Moreover, there are a number of special filler materials available, not generally advertised or discussed because they are considerably more expensive, and their applications are limited.

2. WHAT DETERMINES THE SELECTION OF A FILLER MATERIAL?

Filler material selection is based upon the type of paper, the type of finish required, by roll life requirements, and, sometimes, by cost. This is generally a determination made by the individual mill through suggestions and recommendations from the roll builder.

3. WHY THE DIFFERENT DUROMETER HARDNESS?

Because various nip actions are required to achieve the variety of finishes and/or smoothness desired by printers and converters.

4. HOW IMPORTANT ARE KEYS?

Keys are very important, as they prevent the filler material from slipping circumferentially on the roll shaft. Without them, even the most carefully made roll will be subject to rotation of the filler material on its shaft when a "crash" stop is made with the stack under pressure.

5. HOW IMPORTANT IS THE ELEMENT OF TIME WHEN REFILLING ROLLS?

Since one of the most important qualities of a filled roll is uniform density from end-to-end, such rolls must remain in the hydraulic press for a length of time compatible with the size of roll, pounds of filler, type of filler, and desired density. The time factor is carefully calculated to insure uniform density.

6. WHAT DETERMINES AMOUNT OF PRESSURE USED IN REFILLING ROLLS?

The amount of pressure utilized in the press is determined by the same factors that apply to "time in press".

D. GENERAL QUESTIONS

1. WHAT IS THE SUGGESTED "RUN-IN" PROCEDURE?

There is no universal "run-in" procedure. Each mill and supercalender operator knows best how to condition his specific filled rolls.

As a roll builder who wishes to protect his customer and his product to the fullest extent, we would like to see certain procedures followed. However, we, at the same time, recognize full well that downtime on a stack must be kept to a minimum, and only the individual mill can determine whether it can afford to follow a procedure that, obviously, involves a time factor.

2. IS HAND SANDING NECESSARY?

Hand sanding is not generally recommended to remove a "hot spot". The roll should be removed from the stack and resurfaced in a lathe or roll grinder. If hand sanding is the chosen method, it should be done with a circular grinder!

3. WHAT IS THE BEST METHOD OF CHECKING ROLL HARDNESS?

For years, the Shore Instrument Company Durometer, with "D" scale, was the only recognized measure of hardness for supercalender rolls. In more recent years, attempts have been made to adapt the Rhometer and the TMI Impact Tester to this application. All of these instruments only measure surface characteristics, are only as reliable as the individual is experienced in its use, and can provide acceptable readings only when the roll is new, or newly resurfaced with all "crust" removed! None truly and accurately identify "density" of the roll; density being the ration of unit weight of filler material to unit volume.

Many roll manufacturers, internally, are now basing their quality control and roll hardness on density rather than on Durometer indications. Nevertheless, mills still find it more convenient to use Shore Durometer "D" designations.

A Shore "D" Durometer instrument, calibrated to an 85 D to show uniform density, rather than actual density, is still the most widely used method. (Any instrument calibrated to a "50 point" test block should be recalibrated by the factory). Furthermore, like many similar instruments, the Durometer produces its most accurate readings in the hands of an experienced person. Uniform hand pressure and proper placement on the surface of the roll are mandatory!

4. WHAT IS "BARRELING" AND WHAT IS THE CAUSE?

"Barreling" is a separation of the mass of filler material from the roll shaft. This can be caused by insufficient interference fit of the material to the shaft when filling, or by excessive pressure in the press. A very important factor, often overlooked in evaluating filled roll problems, is very high operating pressure in the stack on the wrong type of filler material, causing the i.d. of the filler material to elongate, with resultant "barreling".

5. WHAT EFFECT DOES A "BARRELED" ROLL HAVE ON OPERATION?

A uniformly "barrelled" roll has very little effect on the supering operation - on roll life - possibly - but not operation! However, very few rolls ever "barrel" uniformly.

Non-uniform "barreling" produces a non-uniform nip action and non-uniform supering. The roll will <u>not</u> have normal life and quality of finish <u>is</u> affected!

6. CAN A "BARRELED" ROLL BE CORRECTED?

Definitely not! It must be refilled!

7. HOW IMPORTANT IS IT TO KEEP ACCURATE ROLL LIFE RECORDS?

We think it is important that a mill checks the life of a filled roll against the life of other filled rolls. This is a quality check against rolls by the same roll builder, or by different roll builders. Additionally, a permanent record exists which provides the history of each roll, pinpointing any unusual events affecting roll life.

8. HOW DO YOU STORE FILLED ROLLS?

The best method is to store the rolls vertically in racks; but, the more common method is to wrap the rolls and store them horizontally in racks, and <u>rotate</u> each roll 90° once a week.

Thirteen
Latest Winding Trends

G. Forsberg

A. TWO DRUM WINDERS

The two drum winder has been the standard in the paper industry for many years, and is still the common choice for most paper machine and paper finishing applications, especially for the common, high volume commodity grades of paper and board. The one exception has been supercalendered LWC grades, where "duplex type" winders have been increasingly used in recent years.

INCREASING DEMAND FOR BETTER ROLL QUALITY

There are the obvious, visible characteristics of rewound rolls of paper that have always been used by inspectors to classify shipping rolls as acceptable or not acceptable. These are normally divided into two broad classifications:

1. Paper defects, for which the winder is not held responsible, slime holes, wire holes, felt holes, wrinkles, wet and dry end, poor quality, start-up paper, poor formation, off basis weight, off color, water marks, wet streaks, off caliper, etc. scrabs, calender cuts, cockles, etc.

2. Winding defects, for which the winder and/or winder operator is responsible, improper width and diameter, wrong core, core length, or offset cores, poor slitter cuts, slitter dust, poor starts, wrinkles, uneven winding (offsets), dished rolls, pop-opens, bad slices, run-ins (run togethers), etc.

In addition, it is possible for an inspector to "sound" the rolls across their width for hard and soft areas, or use one of several roll hardness measuring devices to do this. Excessive variations in cross machine uniformity can cause any given roll to be culled.

However, it is not easy to detect whether the structure of any roll being inspected is uniform enough from the core to the outside of the roll, to assure that the roll will not "star" or become out-of-round during handling and shipment.

For years the use of automatic rider roll programs, and more recently the use of drum torque programs, has made it possible to produce more consistently uniform shipping roll structure form any given winder.

But the ever-increasing demand for better shipping roll quality in recent years, has resulted in the trend towards:

a) Flexible, easier to change programmable control systems that permit more accurate control of the winding parameters over a wider range of operation, by simple adjustment.

b) Continuous roll density monitoring systems, which provide and immediate read-out of the roll density of each set as it is being wound, as well as a record of the actual winding parameters as they were programmed for that set.

c) Very recently there have been installation of continuous automatic roll density control systems. A sophisticated computer actually controls the rewinding roll density in an effort to match it to an ideal curve. it does this by frequent sampling and corrective adjustment for any deviations from the ideal curve shape.

INCREASING DEMAND FOR HIGHER PRODUCTIVITY

The modern paper machine is wider, and operates at much higher speeds than in the past. The increasing use of computer controls on paper machines has reduced lost production on grade changes, and has otherwise helped to steadily improve machine efficiency.

At the same time, the pressure to minimize inventories and maximize service to the paper users has placed increasing demands on the paper machine winder.

New winders routinely run in the 7000 to 8000 ft/min. range, and have frequent order changes that necessitate moving slitters and changing core sizes.

In order for the winder to keep up with a modern, high speed paper machine, it is imperative that the winder accelerate and decelerate quickly, run a very high average speed, and that down-time between sets and reels are reduced to an absolute minimum.

This has resulted in an inescapable trend towards more winder automation, such as:

a) Pre-set, rapid rates of acceleration to top speed, and automatic initiation of deceleration when the required finished roll diameter (or length of paper in the roll) is approached.

b) Self-threading systems to help the operators thread the tail through the winder.

c) Automatic slitter positioning to eliminate down-time or order changes. This is usually accomplished with a small, special purpose computer.

d) Automatic set change equipment, which relieves the operators of the functions required during a set change, by automatically ejecting the finished set, inserting the new cores, cutting off the tails, attaching the tails to the cores, lowering the rider roll, activating safety guards, etc., and starting up the winder.

INCREASING DEMAND FOR REDUCED WASTE

Tougher competition in the marketplace has forced paper mills to strive for a better quality product at a more competitive price. Improved productivity through wider and faster machines running at higher efficiencies certainly helps reduce unit costs and improve competitiveness. But in many cases costs can also be lowered by simply reduced waste (production losses) at the winder.

To aid in this cost containment effort, the winder cull roll records should be continuously reviewed for clues to help reduce losses. Culls due to paper defects are the responsibility of the paper machine operation, but there is also usually ample room for improvement in reducing culls due to deficiencies in either the winder or its crews.

The following are areas that should receive close attention in an attempt to reduce losses in the winding process:

1. Periodic Evaluation and Upgrading of Winders, combined with a comprehensive preventative maintenance program, is mandatory. Minor and major rebuilds, or even complete winder replacement if conditions dictate, will maintain quality production with acceptable levels of waste. A good preventative maintenance program will help guard against unexpected outages and periodic instances of high cull roll rates and/or customer complaints due to mechanical or control system inadequacies.

   A high rate of return on investment can often be achieved if problem solving techniques can determine the cause of production problems, high cull roll rates, and customer complaints, and the necessary steps taken to prevent these expensive problems before they occur, or at least as soon as they start to occur.

2. More Reliable Slitter Operations are important on winders. At higher winder operating speeds, especially on lighter weight grades, more attention should be focused on the slitters.

   The design of the slitters must assure that the shear angle stays constant, the blade loading is correct, and that vibration of the slitter assemblies is minimized. Slit quality should not be subject to degradation everytime the slitters are repositioned.

   Maintenance of the slitter assemblies, motors, knives, etc. is very important. Proper knife grinding techniques are necessary, and the best possible knife materials can make a big difference in slitter life between grindings.

   At higher production rates, changing slitter setting and changing out knives must be accomplished quickly and easily.

   Occasionally one finds a winder that is speed-limited by an inadequate trim removal system.

   Non-contacting slitters have long been a winderman's dream. Water jet slitters and laser slitting systems have been under development for years, but are still not considered tried-and-proven for higher production winders.

3. More Powerful Spreading Systems can result in far fewer cull rolls if bad starts, run-ins, dished rolls, etc. are problems.

   The simple D-bar spreader found on most old winders is today being augmented or replaced by more effective spreading devices of many types, mostly rotating elements that reduce dusting, and have a powerful steering effect on slit webs of paper.

   Bowed roll spreaders introduced to production winders in the 1950's, with mixed results, are today very effective spreading devices because we have learned how to use them. The correct amount of bow and shape of the bow, the use of venta-grooving and carbide surfaces to enhance traction between the roll and the web, and correct positioning of the bowed rolls, have resulted in spreading systems that are extremely reliable.

   Other specially designed spreading devices have their place as well, and there is no longer any need to live with high losses due to spreading problems.

4. Miscellaneous Items, such as good core quality, matched cores, (the same diameter), straight cores, etc., plus winder electric drives that provide accurate, reliable thread and crawl speeds, "soft" smooth transitions into high rates of acceleration and deceleration, positive stopping, accurate tension control, etc., are all important to winder efficiency.

   Good splicing techniques on the part of the winder crew can also reduce cull rolls and customer complaints. This is a part of operator training, which should be an on-going program in every paper mill.

B. MULTIPLE INDIVIDUAL WINDING STATION WINDERS

Duplex winders have been used for over 30 years for winding of plastic film and foil. About 25 years ago, duplex winders were introduced into the paper industry for special applications like winding cigarette paper, narrow rolls of envelope paper, and eventually coated, supercalendered paper.

Due to the inherent low productivity and complex design of true duplex winders, those used in the paper industry quickly evolved into simpler individual station type winders that wound better quality rolls of LWC grades than was possible from existing two drum winders.

Since the mid-60's both two drum winders and the individual station winders have been improved and developed to overcome their different drawbacks.

Two drum winders were built with large drum diameters having special drum grooving, incorporating sophisticated roll hardness programs, and powerful spreading systems to facilitate running very low unwind tension. This enabled them to produce good quality shipping rolls up to 50" diameter, and up to 90" wide, on standard 3" I.D. cores, providing the paper profiles were good enough. Then came the trend in recent years to more rotogravure grades using latex coatings, and the problems with two drum winders, even the most sophisticated ones, became much more critical.

The individual station winders were inherently so much less productive than two drum winders that all the initial installations used two winders for each paper machine. In recent years they have been so highly automated that in many cases one winder is able to handle the full production from a high speed LWC machine.

But the individual station winders had other problems besides productivity. One of the biggest has been the problems of winding defects near the core, especially with iron cores, that comes from supporting the full weight of each roll of the core as it winds up.

Another big problem has been the chucks that support the cores, especially iron cores. Expandable chucks for iron cores have been less than satisfactory, mostly due to the heavy maintenance problem they represent. The use of conventional solid chucks necessitates conditioning the ends of each returnable iron core to a very close tolerance fit so as to avoid vibration of the rotating core on the chucks at high rpm.

There is also a limitation in the minimum width of roll that can be wound on each individual winding station. There is no such limitation on two drum winders.

Recent developments in individual station winders have attempted to overcome the core support problem by designing the winding stations so that part of the weight of the rewinding roll is supported by the single winder drum. This technique reduces the chuck design problem on 3" I.D. iron cores, and also reduces the tendency towards winding defects near the core that comes from supporting the full weight of each roll on the core.

In summary, the individual station winder is inherently so much more forgiving of reel condition than two drum winders that, especially on rotogravure grades, they are slowly by steadily superseding the two drum winder as the standard of the industry for LWC grades.

Fourteen
The Rationale For A Vari-Top Single Drum Winder At Boise Cascade

S. Hardy

Our Division comprises Paper Machines Nos. 10, 11 and 12, all equipped with on-machine coaters, producing #4 and #5 C2S papers. We make rolls only for web offset and rotogravure printing, in basis weights of 40# - 70#/3,300 s.f. (60 - 102 g/m^2).

Our roll diameters range from 40" to 50". While the rolls for web offset are narrow, sometimes down to 23" or less, those for rotogravure are 70" wide and may go wider in the future.

We recognized several years ago that we needed single drum winders with individual rewind stations to obtain really good roll structures for our papers. We purchased a Jagenberg VARI-ROLL winder in 1979 and were very impressed with its performance. It is giving us rolls considerably superior to those previously wound on two-drum winders.

When we began thinking of a new winder, suitable especially for our very heavy rotogravure rolls, we were pushing the capabilities of our VARI-ROLL to the limit. Rolls of 70 inch width and 50" diameter weigh about 6,000#, more weight than we can expect a station to carry and still provide good performance. We faced problems in the other direction as well, as we needed offset rolls that were only 22" to 23" wide.

After considerable investigation, we purchased and installed a Jagenberg VARI-TOP winder, only the second such machine in this country. This winder takes webs up to 165" wide and is equipped with four heavy duty and seven regular rewind stations. The heavy duty stations can accept loads far beyond our present needs, though we anticipate winding 90" wide rolls, weighing about 8,000# each.

The VARI-TOP supports part of the roll weight on its oversized drum and is programmed for a loading and relieving cycle to even out roll density over the entire diameter. We are winding very satisfactory rolls on this machine, rolls which are not hard near the periphery.

As we do on the VARI-ROLL, we use both fiber and iron cores on the VARI-TOP. We obviously prefer the fiber cores, but our expanding chucks allow us to run either way.

We had anticipated some operating problems with the VARI-TOP, due to the fact that its slitter section is located in the basement. This has not materialized, and we have found that the massive construction of this type of equipment makes it a quiet, smooth running machine.

Fifteen
Midtec's Experience With Winders

D. Wereley

1. Midtec's Winding Equipment

   A. Winders
   The mill has three single drum winders that will handle a wind-up trim of 216". All of these winders were installed since 1980. Two original mill winders of the two-drum type are still in operation, handling a sheet width of 150".

   B. Rewinders
   Since 1980, three high speed rewinders, one a single drum, and two of the two-drum design, were installed, which allowed a total of five obsolete rewinders to be shut down and permanently retired.

2. Features of Midtec's Latest Winders and Rewinders

   A. Winders
   We have two Jagenberg Vari-Roll winders with a maximum speed of 7200 fpm, each with eight stations, etc. and one Appleton single drum used for sheet stock.

3. Operation of a Two-Drum vs. a Single Drum Rewinder
   A review of the features of each type and the operating characteristics is being given.

4. Requirements of Winders for Sheet Rolls and Shipping Rolls
   The difference of the two types of rolls are being explored; the impact of the installation of a single drum winder on our sheeting department is being analyzed.

5. Midtec's Experience on Roll Quality by Replacing a Two-Drum with a Single Drum Winder
   In 1984, a single drum winder replaced two two-drum winders as part of the rebuild of No. 5 Machine. Our experiences on roll quality are explained.

Sixteen
Introduction to Escanaba's Winding Facilities

J. Beach, D. Johnson, & E. Winchester

Part A, No. 1 Mill
Jack Beach
Converting Superintendent
Mead Corp., Publishing Paper Div.
Escanaba, Michigan

1. Production Winder
   Jagenberg, Vari-Roll®, Type 650-15
   Trimmed Width 157.5"
   No. of Rewind Stations 7
   Individually adjustable nip pressures
   Maximum Operating Speed 6,500 fpm
   Maximum Unwind Diameter 106"
   Maximum Rewind Diameter 60"
   Chucks for iron or fiber cores

2. Reprocessing Winder
   Cameron, Two-drum
   Trimmed Width 90"
   Slitters for 3 Rolls
   Maximum Operating Speed 3,500 fpm

3. Reprocessing Winder
   Stanford, Roll Doctor
   Trimmed Width 50"
   Sl;itters for 4 Rolls

Part B, No. 3 Mill
Dick Johnson
Converting Superintendent
Mead Corp., Publishing Paper Div.
Escanaba, Michigan

Production Winders
Two Cameron, M.I.R., Single Drum
Trimmed Width 280"
Oscillating Backstands
No. of Rewind Stations 8 each
Minimum Slit Width, End Stations 14"
Minimum Slit Width, Other Sta. 28-1/2"
Each station has individual hydraulic torque drive
Individually adjustable nip pressures
Maximum Operating Speed 6,000 fpm
Chucks for iron or fiber cores

The two single drum winders were installed in 1969. The single drum winders with all the hydraulics have been high maintenance items. We are presently installing a new modern Jagenberg single drum to replace one of the older winders. We have decided to stay with a single drum winder because of that winder's reputation of providing the customer with a superior roll, and to eliminate interweaving experienced on two-drum winders running wide webs.

Part C, No. 4 Mill
Earl Winchester
Converting Superintendent
Mead Corp., Publishing Paper Div.
Escanaba, Michigan

Production Winders Nos. 41 and 42
Waertsilae, Two-drum, Shaftless
Trimmed Width 298"
Oscillating Backstands
Maximum Unwind Diameter 108"
Slitters for 14 Rolls
Rotating Slitter Table Rolls, 6" dia. sectional, rear roll tension sensing
Mt. Hope Spreader Rolls upstream and downstream of slitters
Guide Plates and Air Nozzles for Threading
Maximum Rewind Diameter 50"
Maximum Operating Speed 7,500 fpm
Drum Diameter 31-1/2"
Front Drums Molybdenum Coated
Rear Drums Spiral Grooved
Rider Rolls, sectional, programmed relief (three optional programs)

Winders Nos. 41 and 42 were purchased as two-drum winders because of their ability to cut 14 rolls across. With the mini-web business, this is a definite plus.

Reprocessing Winder
Lenox 6000, Two-drum, Shaft Type
Trimmed Width 100"
Shaftless Floor Pickup Backstand
Maximum Unwind Diameter 50"
Maximum Rewind Diameter 50"
Slitters for 7 Rolls
Maximum Operating Speed 4,000 fpm

Reprocessing Winder
Cameron, Two-drum, Shaft Type
Trimmed Width 77"
Shaftless Floor Pickup Backstand
Maximum Unwind Diameter 50"
Maximum Rewind Diameter 50"
Slitters for 6 Rolls
Maximum Operating Speed 3,000 fpm

Seventeen
The Automation Of Roll Set Changes On Fast Winders

T. Greiner

Papermaking is a continuous process. Ideally, the final step at the end of the paper machine, slitting and winding the paper to the roll and core sizes specified by the customer, should be equally continuous. Given the complexities of breaking up a continuous stream of paper into individual rolls of finite length, making splices where necessary, and observing all other customer requirements, it is not surprising that the step of slitting and winding always has been and still is a batch operation.

Automation and sequential controls on winder systems make it possible to reduce winder downtime for roll set changes to as little as 40 seconds from winder stop to start. This compares with at least 3-5 minutes without such time-saving devices and frequently makes the difference between the ability to serve a high-speed paper machine with a single winder and the need to find additional winding capacity.

Saying that a paper machine winder must keep up with a paper machine is a misleading statement. When we consider that time must be allowed for changes of slitter settings and core chucks, minor maintenance, cleaning etc., it becomes apparent that the winder must not just keep pace with the paper michine, it must actually stay ahead of it. As paper machines keep getting faster, this task becomes more difficult and requires reasonably sophisticated solutions. This assumes, of course, that we are not satisfied with the idea of using two winders for one paper machine or even three winders for two paper machines. In terms of capital expenditure, space, and crewing costs, that solution could be prohibitively expensive.

LIMITATIONS OF RUN SPEED

The next "simple" way out is to increase the run speed of the winder. Top speeds of 8500 ft/min and even more are not unusual today when paper machines can operate in the area of 3000 ft/min and are being designed for even higher speeds. But the idea of increasing winder run speed has reached the area where the law of diminishing returns operates. This is true especially when rewinding diameters are fairly small, e.g., on newsprint, where roll diameters above 40 in. are rare. This is also true of heavy grades of paper, where rolls build fast and the winder simply does not run long enough at constant speed to give much increased productivity as a result of higher operating speeds.

What, then, are the critical areas to be worked on to get the most productivity from the winder?

1. The rate of acceleration and deceleration.

2. The downtime required for roll set changes at the rewind.

3. The downtime required for reel changes and threading.

4. The downtime required for slitter setup changes.

DOWNTIME FOR ROLL SET CHANGES

Roll set changes occur with frequent and predictable regularity and have become somewhat more time-consuming since winders have changed to shaftless operation. The challenge lies not so much in reducing the time needed to finish and start a new set of rolls but in shifting as much of this operation as possible into the run-time of the winder; in other words, in dovetailing the change into productive run-time, when the crew is not fully occupied anyway.

The principal tasks to be performed for a set change are these:

1. Stop the winder.

2. Retract core chucks and raise rider roll.

3. Lift guards and raise lowering table or cradle.

4. Eject completed set and cut the web.

5. Lower completed set to floor or processing level.

6. Insert new cores.

7. Insert chucks.

8. Fasten tails to the new cores.

9. Lower rider roll and activate guards.

10. Restart the winder.

Individual winder designs may necessitate slight variations of this sequence, but it conveys the general picture. Functions like raising and lowering the rider roll, retracting and inserting chucks, and raising and lowering guards must be performed during downtime, but can be sequenced mechanically or electrically to be as efficient as possible.

CORE HANDLING

The greatest consumer of time and the most lucrative area of automation is core-handling and tail-severing and fastening.

Without any special devices, the design of some winders allows inserting of cores up to certain diameters before the winder is stopped. This can be done by hand and will save downtime. More mechanization is needed when the time factor becomes critical.

A trough for cores can be provided in front of the winder and pivoted mechanically around the front drum. The cores can be inserted into this trough by hand while the machine is running (Fig. 1). As soon as the winder is stopped, the problem consists of inserting these cores so that they will come to lie in the saddle between the drums and make contact with the tails of the web on the side that has to wrap the cores, i.e., the side that does not touch the drums and is wound inside on the finished roll.

1. A trough in front of the winder pivots around the front drum.

With webs having reasonably high tensile strength, this problem can be solved by ejecting the rolls before the web is cut. The set of rolls is brought to a stop just outside the winder and held under tension by the regenerative unwind brake. The end of the cradle touching the winder or a specially designed roll bridge lifts a few inches only to provide a path for the trough to rotate around the front drum and insert the cores under the web before it is cut (Fig. 2). Only then is the web cut off by a serrated knife rotating around the rear drum and the tails stapled or taped to the cores by automatic devices incorporated into the design of the ejector-knife assembly. This system is suitable for strong webs such as linerboard and corrugating medium.

2. Inserting the core under the web.

For weaker webs like newsprint or other printing grades, it is necessary to combine the cutoff with the ejection of the roll, as otherwise the web may tear as the rolls are moved out of the winder. Figure 3 shows how we can accomplish this. The rear drum is perforated, and a vacuum is created in the drum while the roll change takes place. In this manner, the tails can be held in place even after the web is cut off. The trough for the new cores has been relocated in the ejector, which rotates around the rear drum. The cutoff knife is not incorporated in the ejector as on the previous design, but moves vertically between the winder drums. The trough for the cores is built up to form a clamp, which can have three positions: fully open, partially open (as shown in Fig. 3), and clamped shut.

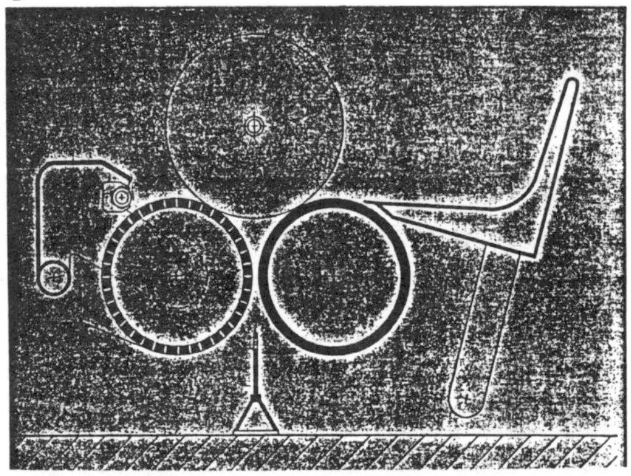

3. Diagram showing the setup for combining cutoff with roll ejection.

As the cores are pushed into the trough, the applicator applies a line of glue to the cores. This requires a special adhesive which retains its tack for a considerable length of time (Fig. 4). On this

4. Applying a line of glue.

application, the entire set of cores is inserted at one time. This is done with a pistonless air cylinder. There are such airy cylinders with strokes up to about 400 in. This means, of course, that there must be space at one side of the winder (usually the drive side) equaling the full width of the machine. It is important that this glue line remain in the same position, meaning that the cores must not rotate within the trough. For this reason, the trough has a clamshell design and closes around the cores as soon as the full set has been inserted. This takes place while the winder is running, and no downtime is needed up to this point.

After the winder is stopped, vacuum holds the sheet to the rear drum. At the same time, the cutoff knife moves up between the drums and its serrated blade flips to make contact with the web at the rear drum, but no actual cutoff takes place until the ejector has moved the roll setup over the front drum (Fig. 5).

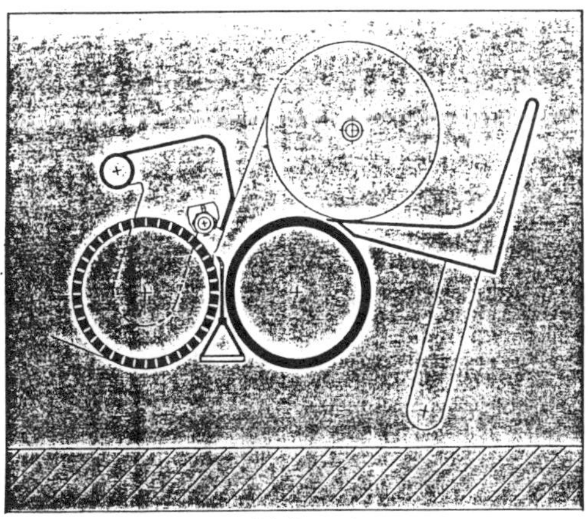

5. Diagram showing cutoff knife moving up between the drums.

Figure 6 shows the moment just prior to cutoff. The web is held against the rear drum by vacuum, and the cutoff knife rests against the web. The core clamp is closed tightly, and the glue line on the cores points down. As the ejector rotates to its final position, the roll set causes the web to break along the serrated edge of the knife and come to rest in the lowering cradle. At the same time, the clamshells of the core clamp open completely and deposit the new cores in the saddle between the drums (Fig. 7).

6. The setup just prior to cutoff.

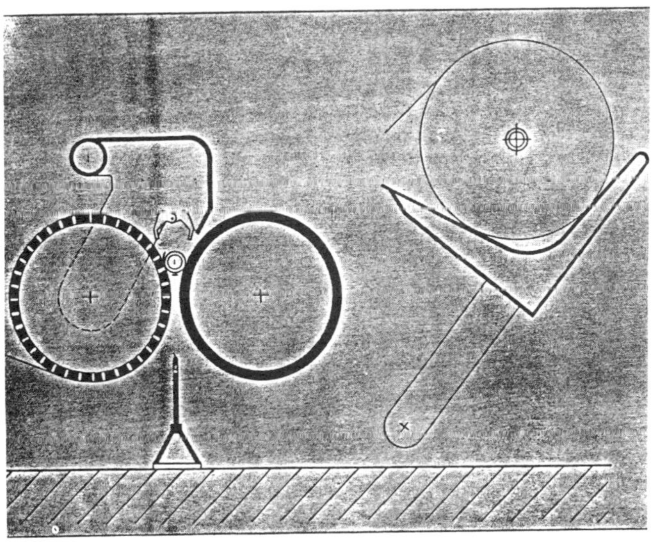

7. The open clamshell depositing a new core.

In Fig. 8, the clamp is open and the cores dropped. The paper is cut off, but does not yet come into contact with the glue line on the cores. This will happen only when the winder starts up again. At that time, the cores will pick up the web after they rotate one-quarter turn, and the tails will wrap the cores as they should. In this system, there is no need for taping or stapling devices to fasten the tails to the cores.

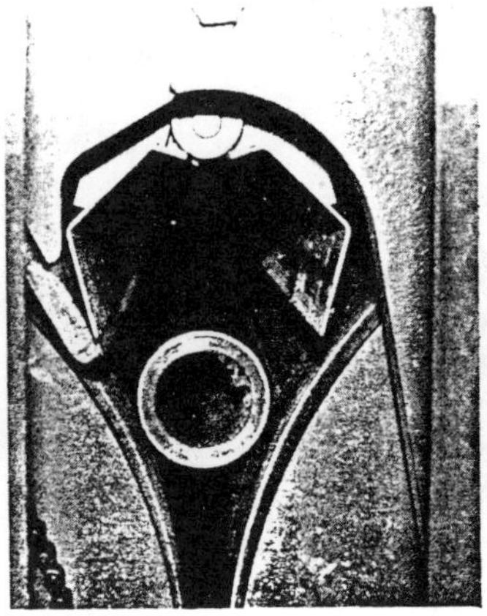

8. The open clamp and dropped core.

In a highly modernized mill with computer-programmed scheduling, the core-handling system can be tied into this schedule without the need for additional personnel. This problem has been solved in and exceptionally fine way by a European newsprint mill on new 335-in. newsprint machine.

The cores are supplied to the paper machine area in lengths equaling the trim of the winder. In this manner, each full set is cut from one length of core and has the same inner diameter and outer diameter. The cores are delivered to the basement, where a circular saw cuts them to the schedule of the particular order. No operator is required. The handling system and the saw are controlled by a microprocessor in accordance with the program of the mill's scheduling system.

These cores are inserted individually, not as the complete set at one time as on the machine described previously. In this second version, it is possible to control the glue application by photo-electric cells in such a way that no glue is applied about 1 in. from either end. This will prevent the adhesive from oozing over the edge of the core (Fig. 9).

9. Photoelectric cells control glue application, keeping glue from the ends of the core.

In this case, the set of cores is assembled in the trough, piece by piece. Once again, the entire operation takes place while the winder is running and requires no downtime. In this version also, the trough is built with clamshells which have an intermediate position between open and closed to adjust to the outer diameter of the cores and permit smooth insertion without giving the cores a chance to rotate.

Since this apparatus requires only a single cut length of core to be inserted at one time, it takes up considerably less space at the side of the winder than the one that inserts all cores with one stroke.

Eighteen
Vari-Step, A Modified Two-Drum Winder For Improved Roll Structure

T. Fernholz

For many decades in the past, the classical two-drum winder with its drums of equal diameter has served as the mainstay of paper mills. Most grades of board and paper, including commodity types of fine paper, have been wound on such winders in a satisfactory manner to rolls of moderate diameters.

By powering each drum with its own DC motor, or by connecting them with a mechanical speed differential drive, and controlling these devices in relation to roll diameter, it was possible to exert some influence on winding hardness. All this under the assumption that finished roll diameters were not excessive and that winder operators were able to make good use of the capabilities of such equipment.

The last few years, however, brought about significant changes both in the requirements of the primary users of paper and the physical properties of certain grades of paper produced by paper mills. Printers receive approximately 40% of the output of paper, worldwide, while 35% is consumed by packaging. Here again, printing is of paramount importance. The trend to higher production speeds and improved quality of printing has led to the demand for better roll structures at increased diameters.

Other factors have contributed to modifications in the properties of many types of paper. Distribution costs have caused basis weights to be lowered while opacity had to be maintained or improved. Fillers and recycled fibers are used to an increasing extent, adding to the density of paper, but affecting adversely its tensile and tear strengths.

Finally, it is necessary to produce rolls with as few splices as possible. Splices impede the ability of the winder to keep up with the paper machine and also mean lost time for the printer or converter. At the same time, the user demands more footage on every roll so as to minimize his need for flying splice changes.

The VARI-STEP® Winder was developed by Jagenberg to meet this demand for roll of mass-produced papers with better structure and larger diameters. This type of winder can fill this need while retaining the operational advantages and practices of the two-drum winder. This applies particularly to papers with moderate apparent densities. A reasonably large range of applications can thus continue to benefit from the efficiency and moderate cost of the two-drum winder.

Fig. 1 Schematic of Vari-Step Winder

The characteristic features of this winder are the two drums of different diameters, mounted on a tilt. This novel machine geometry results in the ability to wind rolls with considerably more uniform density from core to outer rim. This is an essential requirement when rolls of larger diameters are to be wound, and broke is to be reduced.

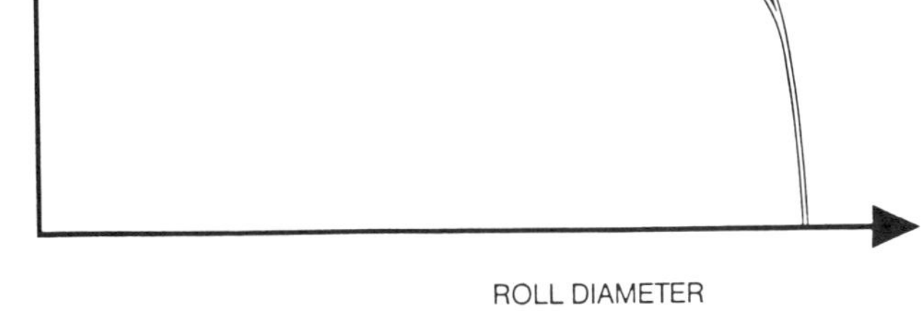

Fig. 2 Ideal Roll Density Diagrams

Optimum roll density means that roll hardness values remain constant or decrease slightly from core to periphery. It is also of importance that values do not vary appreciably from one point of measuring to the next. Peak roll hardness values near the core or the periphery of the roll must not lead to roll defects.

The following parameters, which affect roll density, were analyzed mathematically and by extensive testing:

* Size and geometry of winder drums
* Rider roll nip pressure
* Differential speed of winder drums
* Web tension

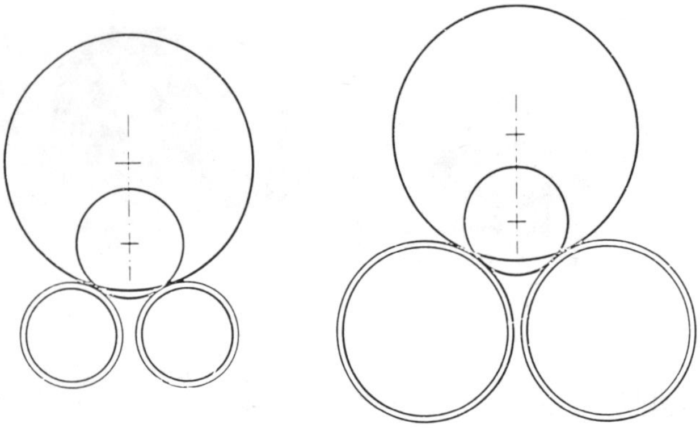

Fig. 3 Small and Large Winder Drums

Roll density (winding hardness) is determined largely by the size and positioning of winder drums. Small diameter drums produce harder rolls than large ones.

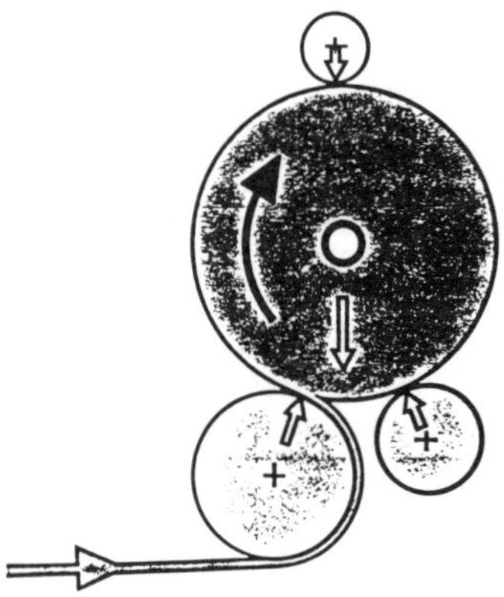

Fig. 4 The Vari-Step is Designed with Drums of Unequal Diameters

As the diameter of the paper roll grows, its weight is shifted to the larger drum.

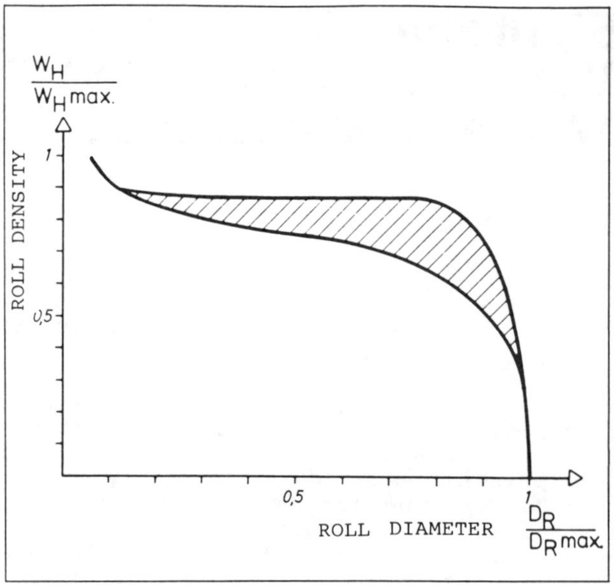

Fig. 5 Roll Hardness Diagram

This gradual shifting of roll weight improves the effect of nip pressures and speed differentials, resulting in a very favorable roll hardness distribution, as demonstrated in Fig. 5. This is a natural process, requiring no action on the part of the operators.

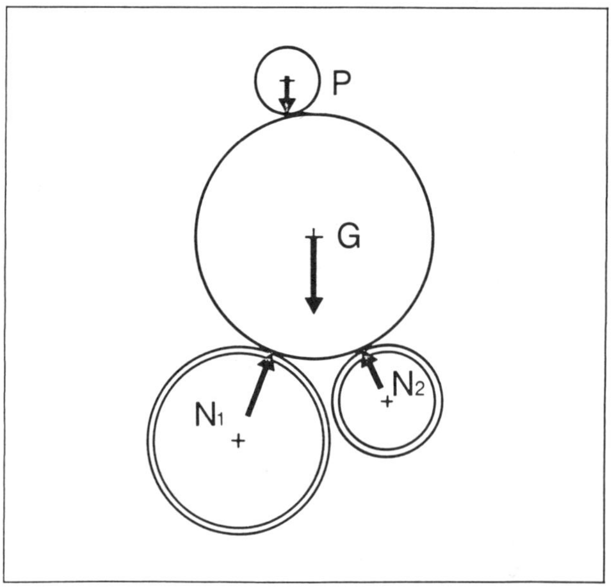

Fig. 6 The Effect of Rider Roll Pressure

In a further effort to even out roll density, load (P) of the rider roll is controlled electronically in such a manner that forces N_1 and N_2 in the nips between drums and paper roll will remain constant over a wide range of diameters. This control is paced by the weight of the roll (G). The nip pressures obtained in this manner produce the optimum level of roll density.

The additional parameters previously mentioned, namely speed differential of the winder drums and web tension, cause the entire density curve to be raised or lowered.

The Vari-Step winder is so designed that it can be powered with a single DC motor drive and a flat belt transmission. In case of extreme grade variations, an adjustable flat belt pulley is mounted on the front drum journal so that the overspeed can be adjusted for very severe changes in grade or basis weight.

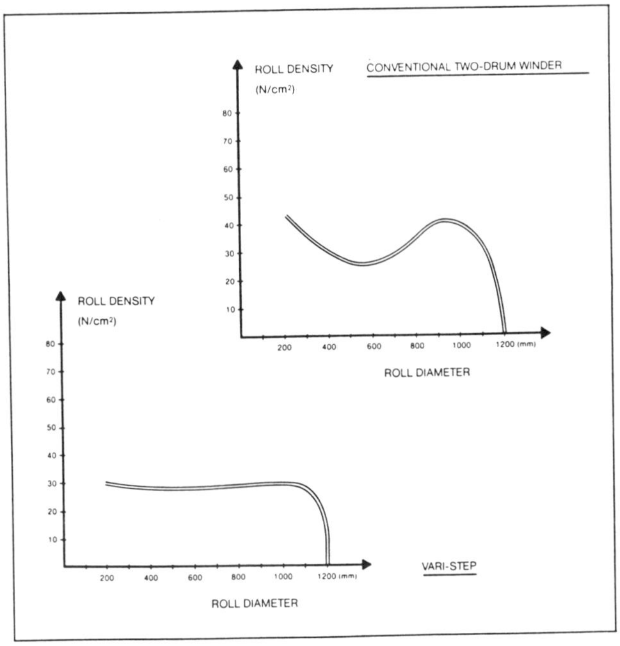

Fig. 7 Contrasting Roll Hardness Diagrams

The upper diagram in Fig. 7 characterizes roll density for a roll wound on a conventional two-drum winder with 20" dia. drums and two-motor speed differential drive. The lower curve depicts the same for a Vari-Step winder with a 30" rear and a 20" front drum. Settings made on the two winders were comparable.

Conclusions

The Vari-Step preserves the operating conveniences of the conventional two-drum winder, while markedly improving roll structure so that larger diameters than previously possible of good quality rolls can be wound.

Nineteen
Modern Equipment For Improving Winder Capacity And Roll Quality

P. Komulainen

The main criteria of the operation of a winder are capacity and roll quality. These levels can be improved by a centralized control system developed for this purpose. The system contains a modern microcomputer-based PC-system that regulates sequential functions, slitter positioning, as well as winding control and measurements. In addition to normal measurements, the system provides for roll density-distribution measurement for evaluating roll quality. Furthermore, the system reduces the risk of breaks at the winder and decreases the quantity of broke and culled rolls.

ROLL QUALITY

The concept of roll quality can be approached by means of a functional analysis of the roll unwinding process. A basic requirement is that a good roll is unwound without disturbances. Disturbances caused by winding include:

* breaks
* wrinkles
* web flutter
* register problems
* dust

The above-mentioned functional properties depend on the winder performance and roll structure. Roll structure can be divided into two parts: roll shape and roll stresses. These are partly interdependent, i.e., defects in internal roll stresses result in shape defects. They are not necessarily observed immediately after winding, but after damage resulting from storing and transporting.

A roll with a good shape has to meet the following requirements:

* dimensions according to specifications
* clean slit
* straight and uniform body
* core in roll center

Requirements concerning stress distribution in a good roll are difficult to present. This results from the fact that measuring roll stresses directly is not practical. However, a density measurement system has been developed (Figure 1) that correlates extremely well with the radial compression pressure of the roll (Fig. 2).

The density measurement works on-line. It has been determined that the stress distribution required in a good roll can be expressed by the density distribution (Fig. 3). In practice, however, the term "roll hardness" is used. Roll hardness must be sufficient near to the core, so that the core will not rotate in relation to the winding roll. This requirement is exaggerated when the roll diameter increases or the outer diameter of the core decreases. On the roll periphery, the roll has to be of sufficient hardness so that the roll remains round in spite of stresses caused by fork-trucks, roll clamps and transportation. Between the core and the outer periphery, the roll hardness must be constant. If the density decreases sharply from the core to the outer periphery, difficulties may arise in a flying splice at the printing machine. Also rewound rolls become harder on the surface than at the core.

CAPACITY OF WINDER

As a concept, the capacity of a winder is complex, since, in addition to the winder itself and its running procedure, capacity is affected by paper and roll properties. Capacity is often evaluated in proportion to the need defined by the paper machine. This can be done by comparing the time needed for making a machine reel on the paper machine with a corresponding average winding time on the winder (Fig. 4).

The capacity of the winder can be reported in many different ways. The expressions "tonnage covered by winder" or "number of sets covered by winder" are generally used. These, however, change as the roll diameter increases (Fig. 5). A better alternative for reporting the capacity is the web length covered by the winder per time unit. This can be directly compared with the running speed of the paper machine. The expression "the equivalent speed of the paper machine" is in fact used here, i.e., the speed at which the paper machine can be run and the number of reels that can go through the winder.

The capacity of the winder is easy to calculate on the basis of given initial values. Table 1 shows an example of such a calculation for newsprint. The data section of the calculation defines those factors upon which winder capacity is dependent.

CONNECTION BETWEEN CAPACITY AND ROLL QUALITY

The capacity of the winder can be increased in many ways. The best way is reducing winder downtime, since it does not have a negative effect on the roll quality. Many other ways for improving the capacity deteriorate the roll quality, such as:

* speed
* acceleration and deceleration rate
* roll diameter
* paper thickness
* parent roll diameter

As paper-machine speeds increase and basis weights decrease, winder operation has to be optimized accurately to obtain both high capacity and good roll quality. It goes without saying that measuring the quality must be possible in connection with such an optimization process. The density measurement provides this possibility.

Wartsila has developed a second-generation control system for winders to optimize both the winder capacity and roll quality. It is a fully integrated control system using a microcomputer-based PC-system (Fig. 6). In this case, the electronic system can automatically handle sequencing of winder functions and slitter positioning. In addition, many gauges can be replaced by color CRT displays (Fig. 7). The first-generation solutions contain a separate logic unit for sequence control as well as microprocessor systems for slitter positioning and winding control (Fig. 8).

A modern winder cannot be operated, if the logic system fails. Therefore, adding more functions to the logic system does not make a winder more vulnerable than it already is. Connections between separate systems are eliminated, and maintenance is simplified.

EQUIPMENT FOR SHORTENING DOWNTIMES

Downtimes that can easiest be shortened are the set-change time and the slitter-positioning time. Set change and slitter positioning were manual functions and thus dependent on crew performance. In addition to shortening these times, automation can reduce manpower requirements and improve safety.

Autochange. Set changes are the most recurring cause of downtime on the winder. With Autochange, set change can be accomplished in 30-40 seconds depending on the technique and the cradle arrangements.

On printing papers, a good solution has proven to be air web threading (Fig. 9). Advantages of this patented system include:

* operating reliability
* elimination of glue

It should be noted that the adherence of the web to the core has nothing to do with core slippage at the printing press. It is essention that initial winds be wrinkle-free and that there is adequate compression between the web and the core.

Automatic Slitter Positioning. When, in addition to continually increasing speeds, it is necessary to reposition slitters frequently, it is essential that the slitter positioning at the winder not consume too much time. Time can be saved by the use of automatic slitter positioning systems (Fig. 10).

These systems permit repositioning to take place during a set change. For this reason, slitter positioning does not affect winder capacity. The time for a slitter pair to be positioned is about two seconds, and the control measurement requires 10 seconds. Total positioning time for a winder with 10 slitter pairs requires only 30 seconds. Accuracy of ± 0.1 mm can be reached in all conditions. An automatic centering for core chucks and the core inserter can also be connected to the system.

SPECIAL FEATURES OF WINDROL SYSTEM

<u>Winding Control</u>. The first capacity-related factor affecting roll quality was speed. Higher running speeds produce softer rolls (Fig. 11). Therefore (and also for stabilizing the web), it is advisable to increase the web tension in proportion to the speed (Fig. 12). The level of compensation can be flexibly adjusted with automatic winder control systems such as Wartsila's WINDROL system.

The acceleration and deceleration rate also affect the roll hardness: the acceleration decreases the roll hardness, and the deceleration increases it (Fig. 13). The level of change depends on the inertia of the roll, i.e., on the roll diameter on a certain paper grade.

A common trend has been to regulate the drum torque-differential which gives a good roll quality at steady running speed, but speed changes are not under control. Another possibility is to adjust the speed difference between the drums. However, this method requires an extremely sophisticated control system plus an extensive knowledge of paper properties and winder mechanics that it is not practical for most applications. A suitable compromise is to regulate the drum current ratio, in which case, for example, the current difference automatically increases during acceleration (Fig. 14). In addition to the regulation of the drum torque ratio, the speed difference of the drums is measured. If the difference is too high, an alarm is sounded.

At the start of a set, the inertia of the roll is low, and the effect of the acceleration rate on the hardness of the roll is of minor importance. Conversely, at the end of a set, the affect of deceleration considerably tightens the surface layers of the roll. This tightening effect increases as roll diameter increases.

When the roll diameter increases 1.25-fold, the inertia of the roll increases 2.5-fold. On a very large diameter roll, the surface layer tends to tighten more than normal simply because of the nip load of the paper on the drums. Therefore, final deceleration of the large diameter rolls should take place over a longer time span than that of small diameter rolls. There are good possibilities for this, as far as the winder capacity is concerned. Figure 4 shows a winding-cycle diagram on 100-cm supercalender (SC) rolls (three sets) and on 125 cm SC rolls (two sets). It can be seen in Table 2 that the average speed of the winder is the same in both cases, although a 2.5-fold deceleration time has been used on the larger roll size. On printing papers, the electric drive of the winder should in fact be such that several acceleration and deceleration rates can be selected (Fig. 15).

It must be possible to choose the acceleration and deceleration rate independently of each other. At the start of winding, it is generally not necessary to use different acceleration rates, since the roll inertia is low. However, it is advisable to use a lower acceleration rate after splicing. If the splice is made on the unwind, the deceleration for splicing has to take place over a longer time frame. With the WINDROL system, the desired acceleration and deceleration rates can be flexibly selected according to the motor power.

As wound roll diameter increases or the paper caliper is reduced, the theoretical winder capacity increases since the web length of the roll grows longer. In this case, the running time of the winder at full speed increases proportionally. However, both an increase in the wound roll diameter and the reduction of paper caliper add to winding difficulties. The level of backstand tension used at the winder must be properly selected in correct proportion to the paper thickness (basis weight) and the roll diameter. A combination of a large roll diameter (a high nip pressure), thin paper and low backstand tension could result in crepe wrinkles. If, by contrast, the running tension is too high, bursts appear in rolls, and the break frequency increases both at the winder and the printing machine (Fig. 16). This can be seen very clearly when running so-called SC jumbo rolls on a two-drum winder.

In the WINDROL system, suitable running tensions, and if necessary, steady running speeds and proper deceleration rates have been recorded for each roll-diameter and paper-thickness combination. In this way, it is easier to find the optimum solution for the roll quality and winder capacity.

The machine-reel size also affects the winder capacity, because the machine-reel change time and web threading of the winder consume more time than a set change.

Another factor is that from the point of view of the winder, present machine-reel winding systems are not adequately developed to produce a good machine reel. Density measurements repeatedly show that sets wound from the machine-reel surface are looser at the start than are following sets (Fig. 17). This tendency increases with reel size.

In this case, for example, a higher rider-roll loading for the first set should be used at the start of a fresh multi-set parent roll (Fig. 18).

Automatic Stopping. Usually wound rolls are ordered according to a certain diameter. The printer usually wants as large a diameter roll as possible. However, they must not be so large that they cannot rotate on the roll stand. Deviations in diameter from those ordered are not desired. However, a comparatively common request is that no splice be made 20 mm or closer to the roll surface. In such cases, an undersize roll is preferred.

The lifting of rolls stacked one above the other places perhaps the greatest demand on the uniformity of a roll's diameter. When the clamps of a lift truck grasp two rolls (Fig. 19), the roll with a larger diameter is crushed. The roll diameter should be as constant as possible so that this is avoided.

Today many winders are stopped semi-automatically. This requires setting the footage counter based on experience. The winder then signals an alarm at the set point, and the operator stops the winder. The winder can also be switched to stop automatically. Inductive diameter measurement devices have also been used to automatically stop the winder.

These systems, however, are not very accurate, since footage and diameters depend on:

* running speed
* deceleration rate
* paper thickness
* roll hardness
* desired final diameter

If the diameter must be accurate, a low running speed must often be used at the end of a set (Fig. 20). This reduces the winder capacity, since the winding cycle is extended. In addition, every additional speed change deteriorates roll structure thus adversely affecting wound roll quality.

Diameter differences may also occur because a machine reel may not be sufficient to fill the last set. If, by contrast, the machine contains too much paper, it unnecessarily adds to the quantity of broke produced. An exact correct amount of paper should in fact be run on the machine reel.

Since the accuracy of today's automatic stopping systems is inadequate, set changes in spite of mechanical automation, are not fully automatic. Similarly, the start-up of the winder after a set change usually takes place manually.

The correct starting moment of deceleration can be calculated by microprocessors on the basis of measurement data. Since final-diameter forecasts, on the basis of which necessary changes are made, are also supplied during the entire deceleration stage, a final-diameter accuracy of $\pm$ 1 mm is obtained in all conditions.

The automatic stopping system requires no extra measurements, since the density-distribution measurement already requires an accurate measurement of the roll diameter and the web length. At the same time, an accurate paper thickness inside the roll is obtained.

The system can also pre-calculate the diameter and the quantity required of a machine reel. the calculated value is accurate, because the web thickness is measured both on the unwind and the windup. When the winding of each set starts, the system forecasts the roll diameter, if the machine reel is evenly divided between the sets.

Automatic stopping is also used for splices to be made at the winder. This can take place, for example, so that the operator measures the paper on the reel spool up to the splicing point. When this data is fed into the system, it stops the winder on the basis of the diameter measurement.

A combination of automatic stopping and an automatic set-change system make fully automatic winding after web threading possible. In other words, the winder receives an order to wind a desired number of sets from a machine reel, after which the winder accelerates and decelerates itself as well as performs the necessary set changes. If a break occurs at the winder, the operation stops automatically.

MEASUREMENTS

The system is designed to measure, for example, the following factors:

Pulse quantities and speed differences

* front/rear drum
* core chuck, Tending Side/Drive Side
* reel
* bottom slitters

Forces and force differences

* tension, TS/DS
* rider-roll load, TS/DS

Currents and current ratio

* front/rear drum

The following machine reel or set quantities can be calculated on the basis of the measurements above:

* diameter
* web length
* average paper thickness
* average roll density
* radial density distribution

When measurement data are printed on a normal plotter, diagrams are in general unfortunately obtained as a function of time. However, most factors are processed as a function of roll diameter. In the WINDROL system, the display can be chosen either as a function of time or roll diameter.

For example, speed and tension can best be presented on the same display as a function of time (Fig. 12), whereas rider-roll load, current ratio and density distribution are shown as a function of diameter (Fig. 21).

To clarify the reasons for breaks, the display screen can also provide a report of changes that have occurred during the last seconds.

The measurement data and the quantities calculated on the basis of them can be provided with alarm limits, in which case the readout concerned starts to blink on the screen (or change color).

RECORDING OF DATA

The end product of paper manufacturing is not paper in general, but a paper roll in particular. Therefore, the quality control and recording of data concerning individual machine reels should be gradually set aside, and the recording of data should focus on individual rolls. This is also defended by the fact that complaints from customers relate to individual rolls.

The WINDROL system can function as a sub-station of a centralized mill computer, and in addition to the above-mentioned measurement data, it can transmit data about the running method of the winder. Similarly, the system accepts data from the mill computer, which reduces the need for data input at the winder.

LITERATURE CITED

Eriksson L.G., Lydig C., Viglund J.A. and Komulainen P., Measurement Of Paper Roll Density During Winding. TAPPI 66(1): 63 (1983).

```
WINDER   CAPACITY   CALCULATION

WÄRTSILÄ  JÄRVENPÄÄ              1.11.1983 PEKKA KOMULAINEN
CUSTOMER X                        PAPER MACHINE 4
TYPE OF WINDER: KL 1000 V+SLITPOSIT

DATA
***********************************************************
BASIS WEIGHT    PM POPE 45      WINDER  45   G/SQ.M
ROLL DENSITY    UNWIND  640     WIND-UP 670  KG/CU.M
WEB WIDTH       PM POPE 8750    TRIMMED 8700 MM
***********************************************************
REEL DIA        MAX     2600    SPOOL   820  MM
WIND-UP DIA     ROLL    1000    CORE OS 100  MM
***********************************************************
RUNNING SPEED   WINDER  1950    PM      1200  MPM
SPEED CHANGE    ACCEL   .6      DECEL  .6 M/S2
WASTE           WEB     1 %     PM BREAKS 1  %
***********************************************************
WIND-UP SPLICE   12 MIN         PER 200 KM      8.5 PCS/D
UNWIND SPLICE    4 MIN          PER 200 KM      8.5 PCS/D
TRIM CHANGE      .5 MIN         PER 200 KM      8.5 PCS/D
***********************************************************
OPERATOR &MECH. EFFICIENCY       94  %
DOWNTIME BETWEEN SETS            35 S
REEL INSERTION TIME              1.5 MIN
WEB THREADING TIME               2 MIN
UNWIND WEB WIDTH                 8750 MM

RESULTS
***********************************************************
WINDER CAPACITY                  663 T/D        146 SETS/D
PAPER MACHINE                    674 T/D        146 SETS/D
***********************************************************
PARENT REEL                     DIA 2420 MM      5 SETS
WEB LENGTH                      SET 11577 M     REEL 57884 M
WEB WEIGHT                      SET 4532 KG     REEL 22792 KG
WEB THICKNESS                   SET 67          REEL 70 MICRONS

BREAK UP OF PARENT REEL CYCLE ON WINDER:
***********************************************************
                  S/SET         S/PARENT REEL    %
***********************************************************
ACCELERATION      54             286             10
CONSTANT SPEED    302            1510            51
DECELERATION      54             286             10
SET CHANGE        35             140             5
REEL CHANGE                      245             8
TRIM CHANGES                     0               0
SPLICES                          309             10
OPERATOR &MAINTENANCE            177             6
***********************************************************
REEL WINDING TIME                2953 S  =       49 MIN 13 S
PM REELING TIME                  2953 S  =       49 MIN 13 S
REMAINDER TIME                   0  S  =         0 % OF PM TIME
SET WINDING TIME                 410 S  =        6 MIN 50 S
```

Appendix 1: Example of winder capacity calculation

	100 cm	125 cm
Acceleration	3 x 60 = 180	2 x 60 = 120
Running at 30 mps	3 x 407 = 1221	2 x 627 = 1254
Deceleration	3 x 60 = 180	2 x 150 = 300
Set-change	2 x 30 = 60	1 x 30 = 30
Machine-reel change	240	240
Total machine-reel time, s	1880	1944
Web length on machine reel, m	41987	43896
Average speed, mps	22.3	22.6

App. 2. Effect of deceleration rate and number of sets on average winder speed

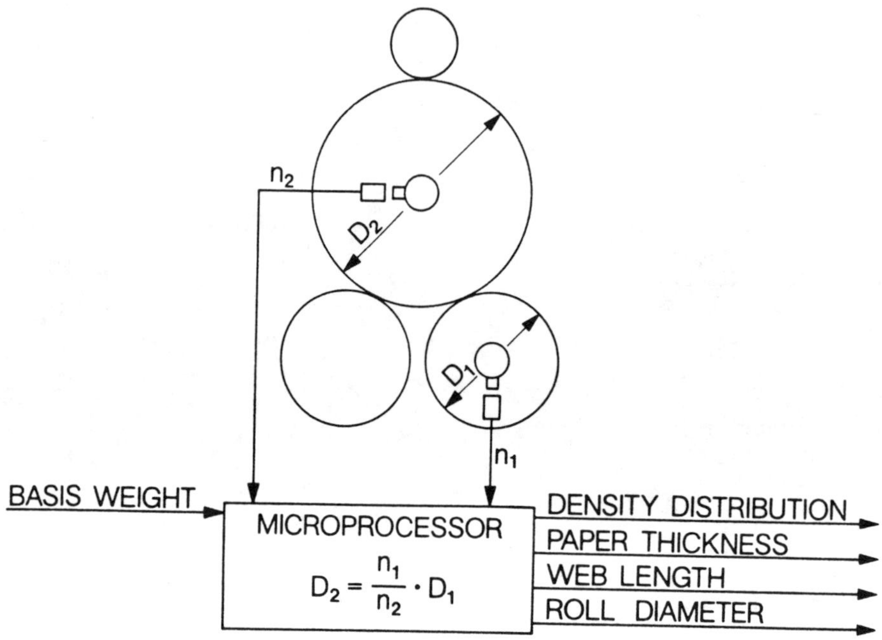

Fig. 1 Principle of paper-roll density measurement

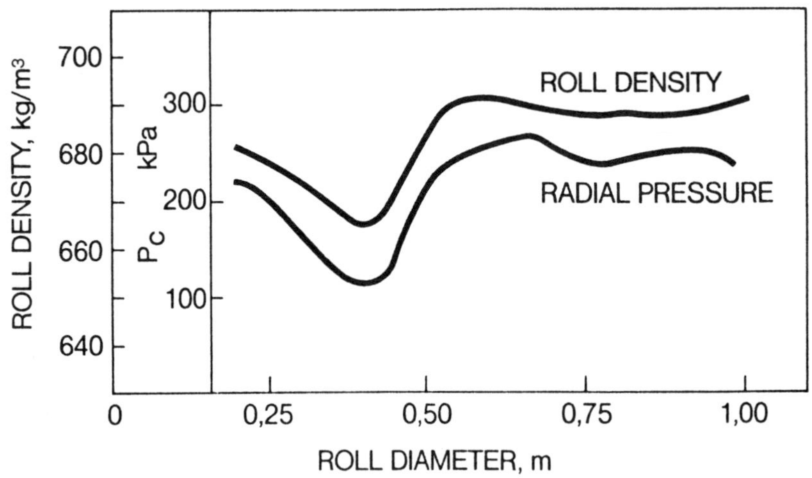

Fig. 2 Relation between roll density and compression pressure

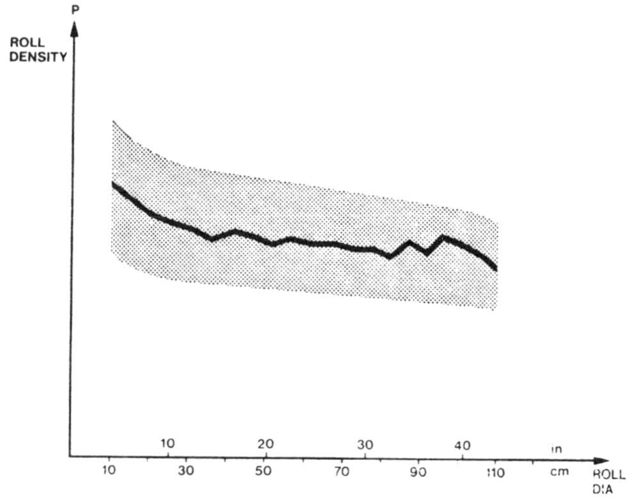

Fig 3. Density distribution of well-structed roll

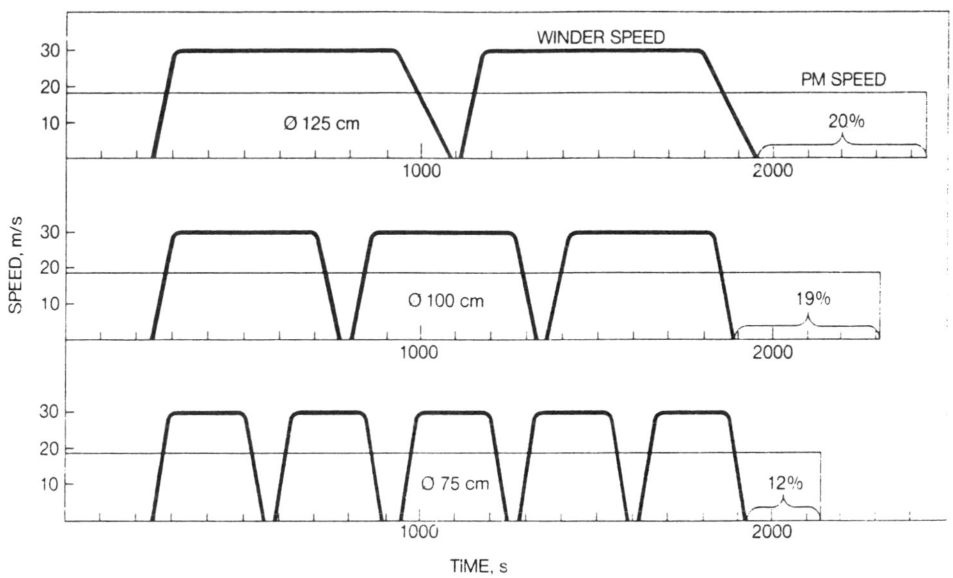

Fig. 4 Winding cycle diagram on different-sized rolls

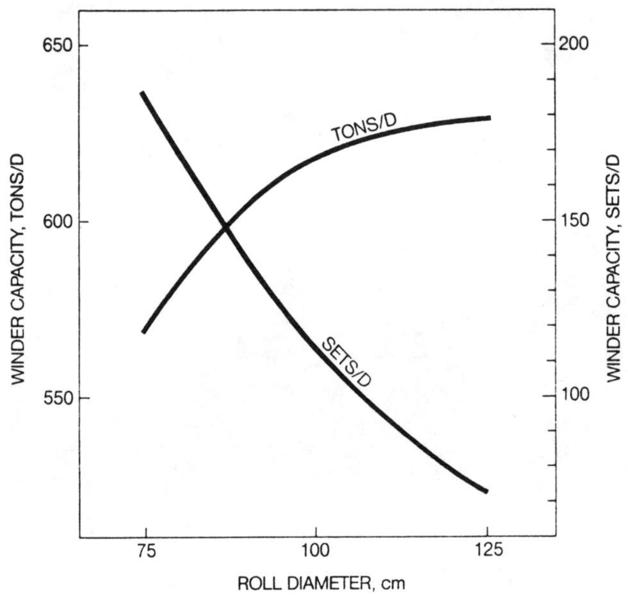

Fig. 5 Winder capacity in tons and as number of sets

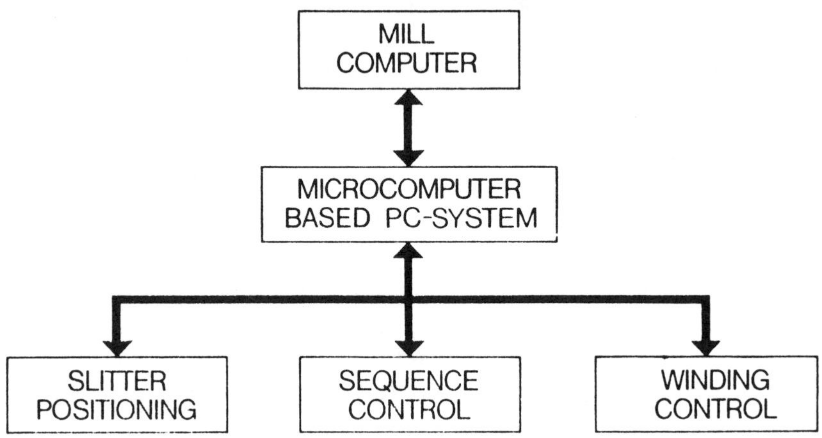

Fig. 6 Centralized winding control system

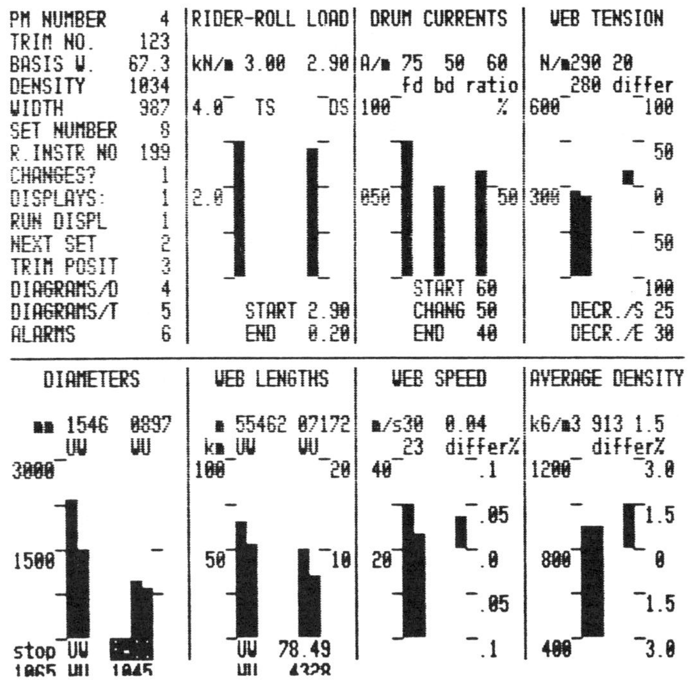

Fig. 7 Example of hard-copy printout

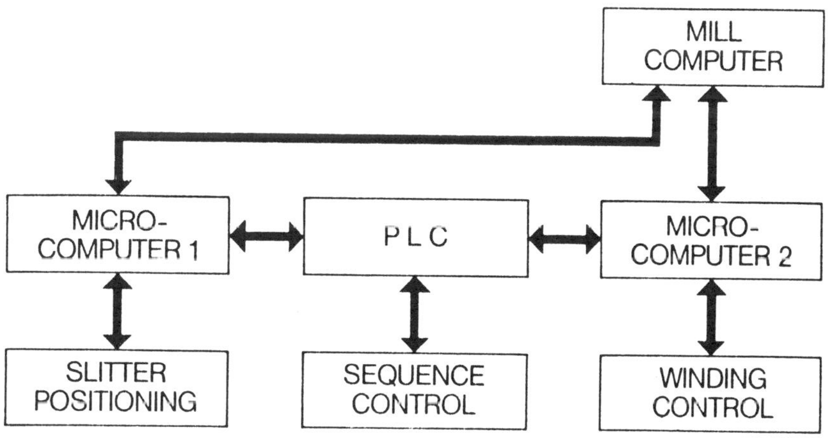

Fig. 8 First-generation system

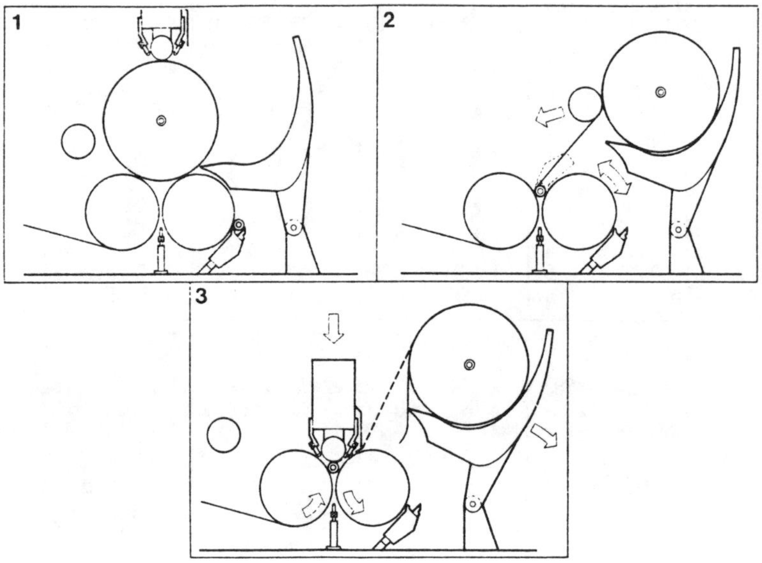

Fig. 9 Automatic set change with air-jet feeding

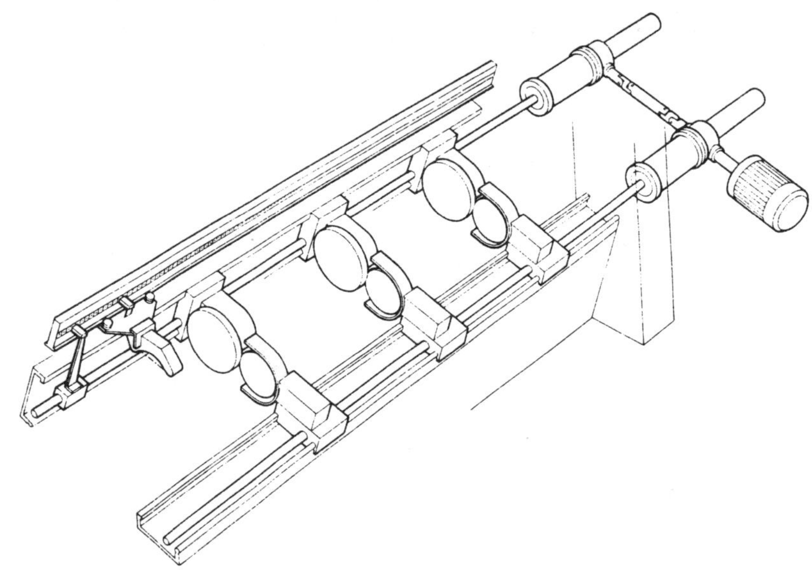

Fig. 10 Automatic slitter positioning system with feedback control

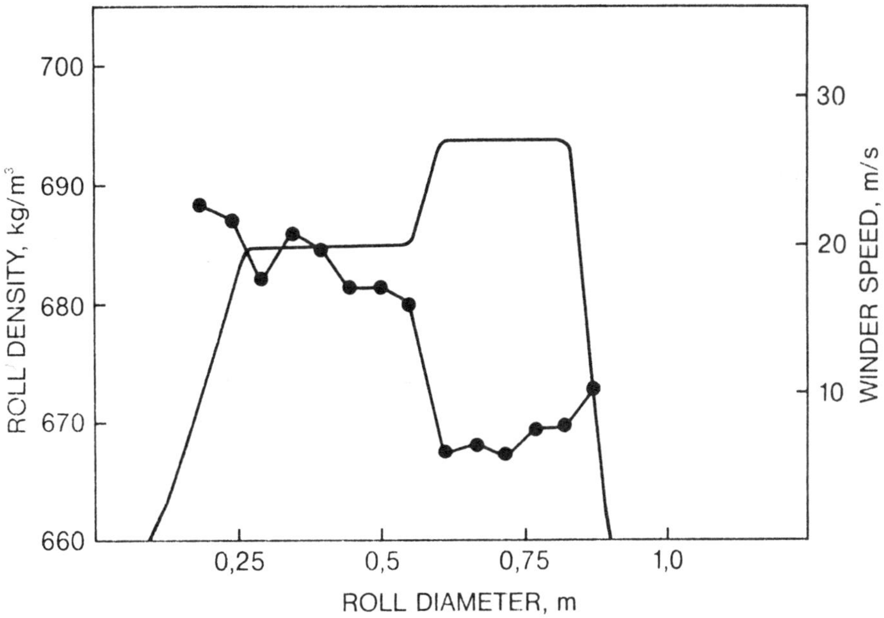

Fig. 11 Effect of running speed on roll density

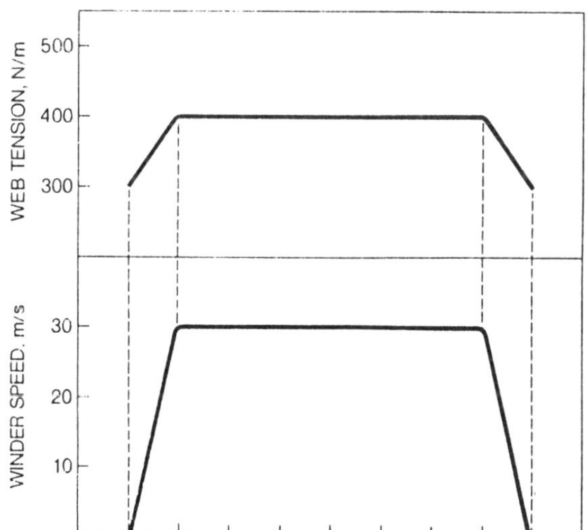

Fig. 12 Elimination of effect of speed through tension

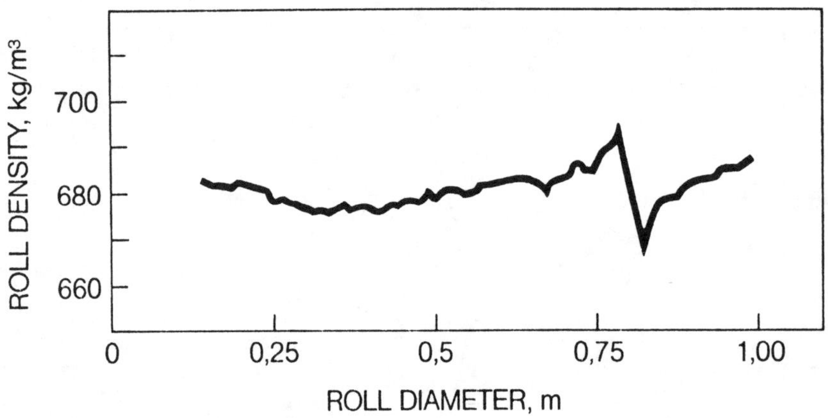

Fig. 13 Effect of splicing on roll density (deceleration and acceleration)

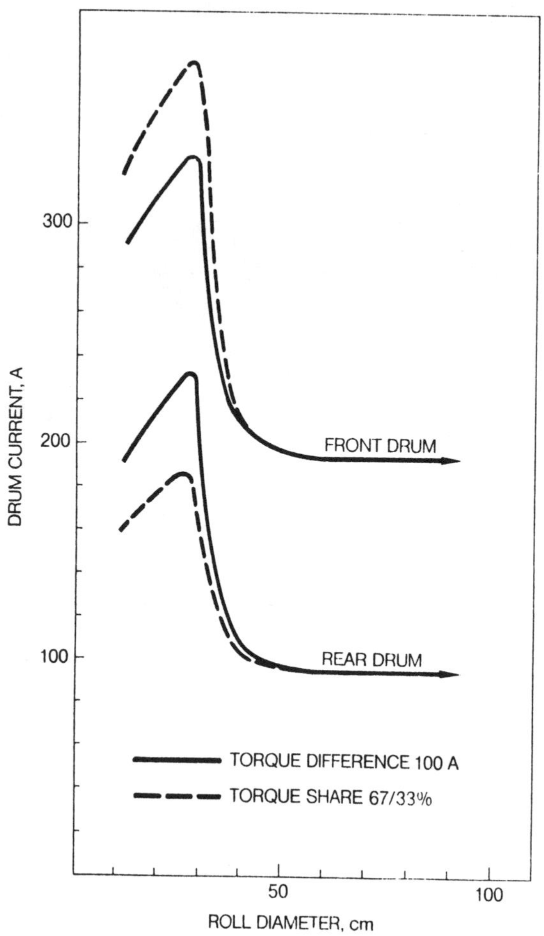

Fig. 14 Drum current-difference and current-ratio control

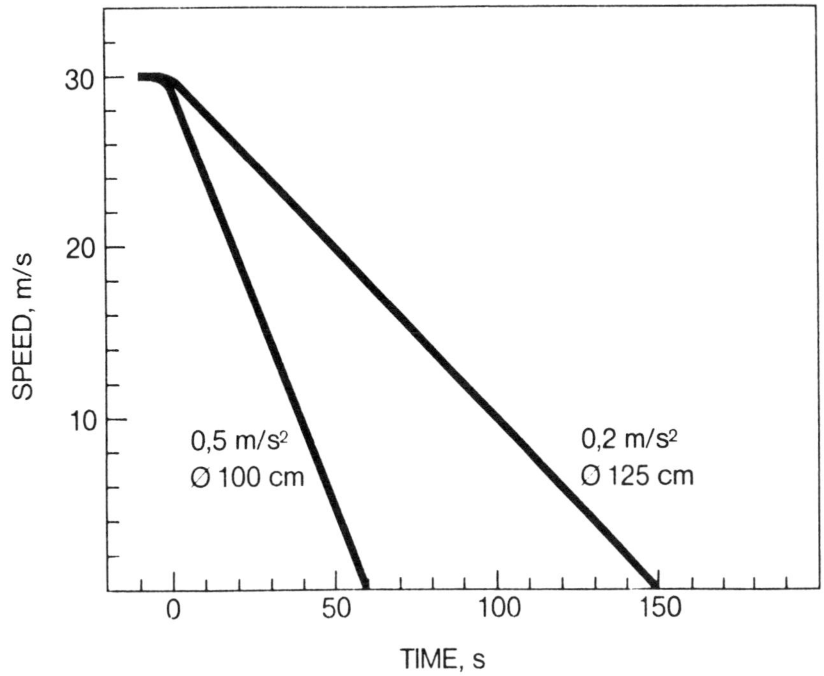

Fig. 15 Use of different deceleration rates on different-sized rolls

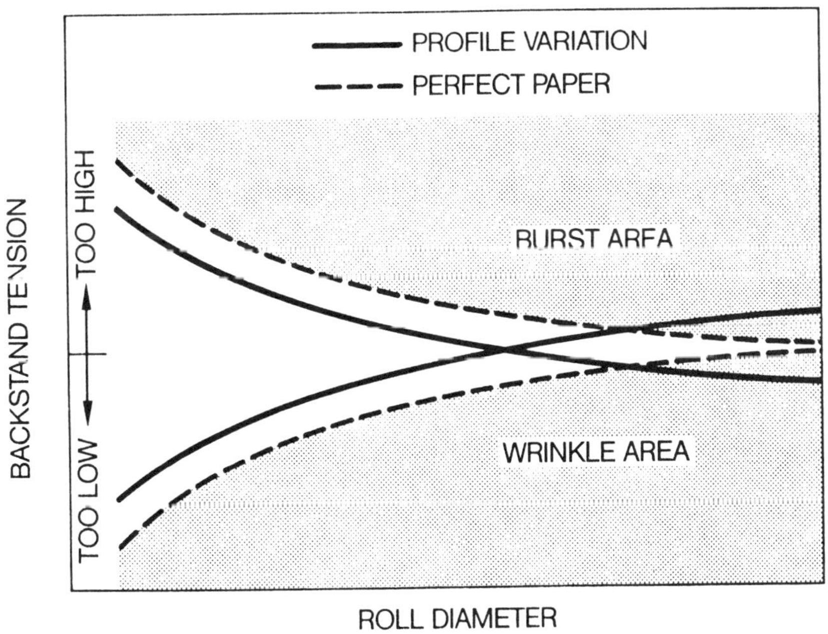

fig. 16 Factors affecting the appearance of crepe wrinkles and bursts

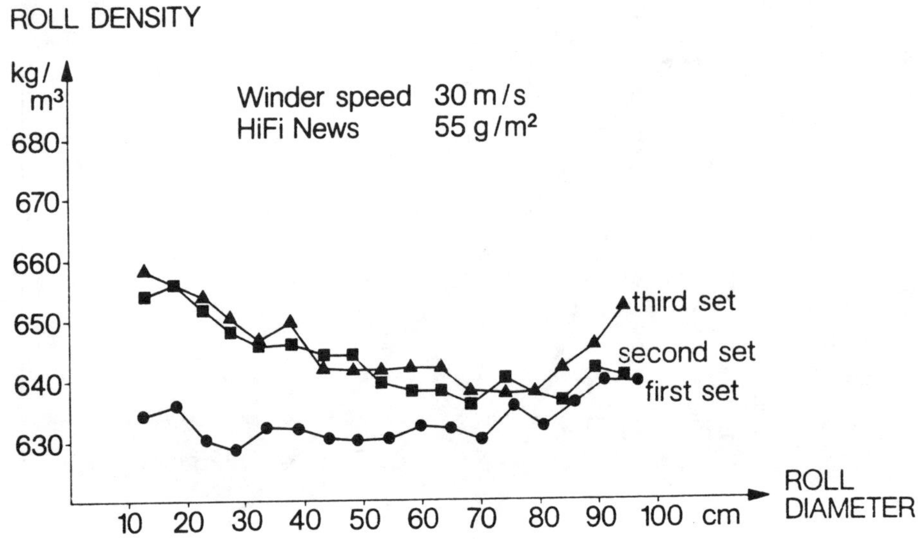

Fig. 17 Density-distribution of a roll wound from the machine-reel surface in comparison with other rolls

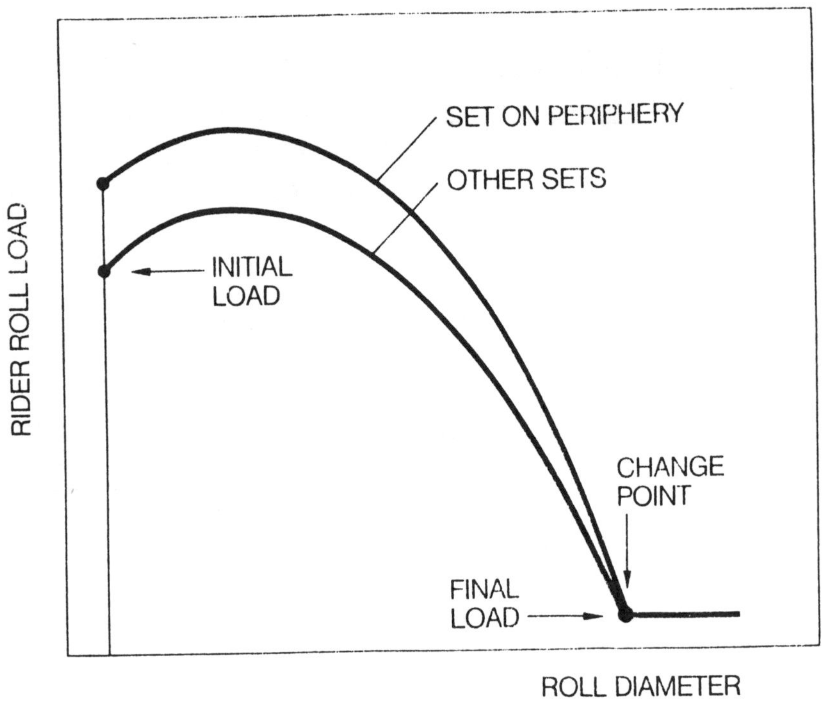

Fig. 18 Compensation of softness of first set (on roll periphery) with rider-roll load

Fig. 19 Effect of roll-diameter difference in handling rolls with clamps of fork truck

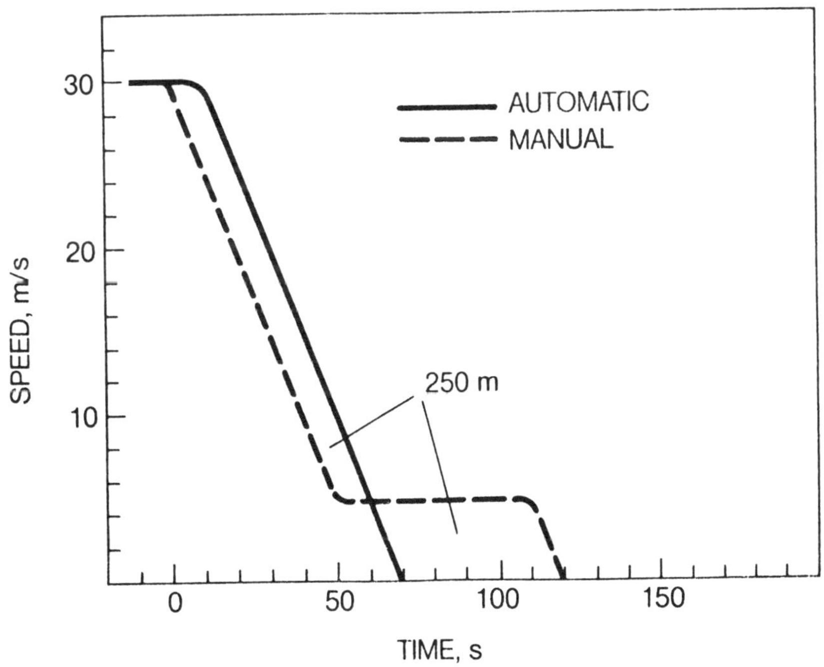

Fig. 20 Running to a desired diameter with manual and automatic control

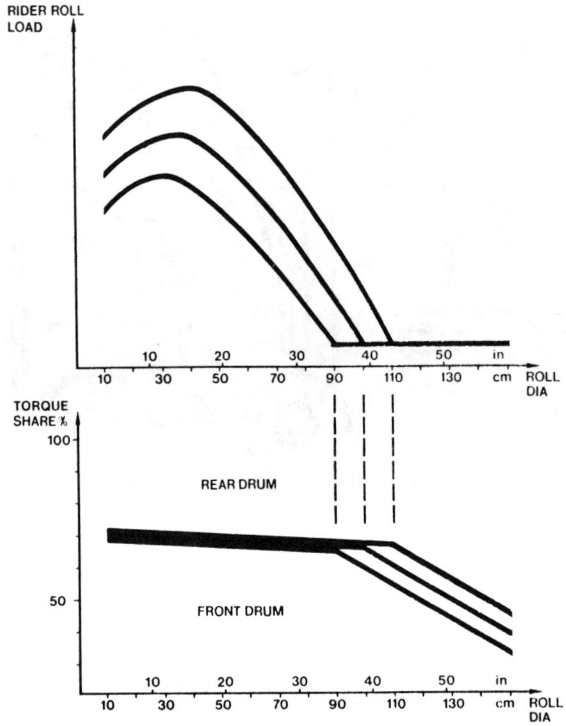

Fig. 21 Programs for rider roll load and current share as a function of roll diameter

Twenty
Rebuilding Machine Winders-Is It Worth It?

R. Bender

The decision to rebuild in place of purchasing new machinery is not as simple as flipping a coin, or turning a card!

What do you expect to gain?

How much can you afford to spend?

Is the existing machine - in your opinion - <u>and</u> in the opinion of outside experts - of an age and condition to warrant expenditures necessary to attain your goals?!

At one time, the feeling existed that a suitable rebuild of any machine could only be done by the original manufacturer of the unit - especially true in the case of winders. Today there are specialists in all areas of machinery rebuilding - people who have the technical know-how to adapt designs used in their own products to existing machines produced by others. They have the sales, design and service personnel, <u>and</u> the manufacturing facilities to convert existing operating machines into "like new" units at costs not exceeding 50% of the cost of a new one -contingent, of course, upon the extent of the rebuild.

Case histories speak, eloquently, for themselves -especially where somewhat unusual circumstances contribute to unusual solutions to problems encountered.

<u>Case #1</u>

In early 1979, Fraser Paper Company, Madawaska, Maine selected the Appleton Machine Company to rebuild a 217 inch face two drum winder. The single most important stipulation was that the successful bidder must provide the very latest in winding technology in this rebuild project. The winder must have the capability of shaftless winding of 60 inch diameter rolls of 24.5#/3000 sq. ft. catalog stock to 42.5#/3000 sq. ft. bond grades at speeds to 5200 fpm -on 3 inch I.D. metal cores.

In order to conform to the customer's strict requirements of state-of-the-art performance on this order, and to insure that there would be no surprises at startup, the entire winder was dismantled and shipped to Appleton, Wisconsin where it was carefully inspected for wear and other ravages of time and continuous use, necessary measurements were taken, and it was reassembled in Appleton's plant.

The decision to dismantle and ship the complete winder to Appleton was a somewhat unusual; but very practical one on the part of the mill; because the winder had been out of service for years. It was not even being used as a standby unit; because a new winder replaced it on #6 paper machine. However, the rebuilding of #3 paper machine created a need for additional winding capacity. Rebuilding and relocating the Cameron unit to #3 machine was a readily justifiable expenditure as compared with purchase of a new winder at that time. The machine arrived (without D.C. electric drum drive) in Appleton's plant in the middle of May, 1979. It was rebuilt, transported back to the mill March 31, 1980, reinstalled, and started-up the first of May, 1980.

Basically, the rebuild consists of the following:

1. Modifications to the oscillating mechanism and controls were required to put the unwind once again in acceptable operating condition, and the stands and saddles were fitted with parts to adapt to spools from #3 and #7 machines.

2. A completely new JET-SET^tm slitting station with vertical web run, mounted on the back of the winder frames, replaces the original slitter section located under the fron drum.

The fifteen (15) state-of-the-art units are carriage mounted on anti-friction ways and located in a manner that provides for easy accessibility. Manual positioning is facilitated by a unique system involving locking top and bottom slitter units together - an arrangement that maintains the preset side loading pressure while permitting each set to be repositioned in 15 seconds. This system eliminates 50% of the hoses required by other systems. Slitters are mechanically locked in place during operation and are pneumatically unlocked. Consequently, slitters do not unlock and "float" out of position if mill air supply is lost to any, or all units.

Bottom slitter bands are individually mounted on D.C. motors whose supports are designed for quick and easy removal. Top slitters are pneumatically loaded into the web run and against the bottom bands.

3. In order to induce the web to flatten and to align itself properly before entering the slitter section, a larger diameter, smooth face, individually motor driven carrying roll is mounted on framework above the slitters.

A pneumatic tube-type web holder can be activiated by the operator to press against the roll and, thus, hold the web in the slitter section when necessary to cull paper from the parent roll, or to splice a tail.

Mounted after the slitter section, in a location selected to provide a constant wrap on the first bowed spreader roll, is a second, large diameter, smooth face, individually motor driven carrying roll. These two large diameter rolls are intended to stabilize the web; at critical points in the web path.

4. The roll located immediately after the slitters (and ahead of the second carrying roll) is a small diameter, low inertia, web driven unit with a sectional aluminum body. The diameter of this roll is such that it always operates under the half-critical, which adversely affect web tension sensing systems, are, thus, avoided. This is of extreme importance, expecially in this instance, because this roll is supported on a set of electric load cells that must accurately sense web tension requirements and transmit resultant information to the controls of the unwind regenerative braking motor. The accuracy of this system has a great bearing upon quality of slit and finished sets.

5. Since the winder is required to rewind a fairly wide range of basis weights in several grades from three (3) machines, and to slit a variety of widths in each set, as well as from set to set, a dual spreader system is mounted after the slitters. These fixed bow spreader rolls function as a set, one roll complementing the other, to uniformly separate and hold slits before the web contacts the first winding drum. Sets are virtually free of stuck rolls, as long as the bowed rolls are in adjustment.

6. A web threading system conveys the tail from the slitter section, under the second carrying roll, through the dual speaders and up between the drums where it can be attached to the cores.

7. Because of the original location of the slitters, the drum height was inconvenient and awkward for operators. The bottom sections of the framework were replaced with heavy fabricated steel pieces that reduce drum height; thus, making the drum area more accessible and providing greater operator convenience.

8. Since one of the major goals of this rebuild was shaftless winding (i.e., total elimination of the old cantilevered coreshaft boxes, slides, raising and lowering and counterbalancing mechanisms) - the coreshaft and rider roll slideways on the existing vertical columns were given a clean-up cut on a planer, and additional vertical columns installed parallel to the existing columns so that core chucking mechanisms could be restrained between two vertical surfaces rather than cantilevered from single columns as before.

(On Cameron winders of this vintage, if the slideways are in reasonably good condition and the frames are not warped, the expeditious and economical approach is to design the carriages for the chucking mechanisms so that one can take advantage of the existing vertical ways for the guiding the anti-friction rollers of the carriages.)

Shaftless winding, through the elimination of the coreshaft, allows each individual roll of a set to find its own position in the "winding pocket". It is no longer restrained by the coreshaft, and, as a result, more uniformly would rolls are produced.

Winding cycles are reduced, as is operator fatigue, since coreshafts do not have to be handled. In addition, the maintenance costs associated with coreshafts and their handling mechanisms are eliminated.

9. The rider roll controls are programmed for fully automatic, or, alternately, manual control. The system allows the operator to set a pre-determined rider roll pressure at the start of the winding cycle and, also, to select a diameter at which rider roll pressure will be reduced to zero. In the automatic mode, the system will taper the pressure from starting pressure (maximum) to zero at a specific diameter.

Such system assists in building uniform density into the wound rolls as roll diameter build up, and allows an operator to duplicate previous settings that gave good results.

Where web, or reel conditions dictate, the operator has the option of total manual control during the cycle.

10. The D.C. drum drive had been a single motor type using speed control only. It was replaced with a two motor torque control drive system through the simple expedient of adding a second motor and new controls.

Combining the new web run, which now wraps the back drum, with drum drive torque control, with closely controlled web tension and with programmed rider roll pressure, provides control of winding roll density equal to any new winder.

The operator has virtually unlimited flexibility in his controls. High torque on the front drum, combined with high rider roll pressure at the start of the winding cycle, produces tight wraps around the cores which is, of course, a prime requirement for winding good rolls. A poor start negates even the most sophisticated controls farther into the winding cycle.

Reduction in front drum torque during the winding cycle assists in controlling wound-in tension produced by nip action between the drums; and by the ever increasing diameter and weight of the winding roll.

These controls function to reduce the amount of internal bursts and potential "starring" of the finished set.

In order to further enhance the effectiveness of these control systems, the winder drums were resurfaced and regrooved with a spiral pattern that assists in evacuation of air that winds into the roll with high speed webs.

11. The last major component of this rebuild is the roll lowering cradle which was added to the front of the winder to lower wound rolls to floor level and to act as a barrier for the winding roll.

The hydraulic system is designed to lower - or raise - a full diameter set. This feature permits the operator to stop the lowering cycle and to return a full roll of paper to drum height, from any position, if there should be any obstacle in the way of completing the lowering cycle. This avoids the necessity to use a crane when such circumstance arises.

An auxiliary control panel is mounted on the front of the winder containing the only controls available to the operator for the cradle. It is designed and located to provide an unobstructed view of the area around the cradle, for safety purposes.

12. All other operator controls necessary to build quality rolls are mounted in a new control console equipped with the latest state-of-the-art control devices.

Abandonment of the original operators' benchboard was necessitated by the obsolescence of many of the manually operated hydraulic and pneumatic controls. New, modern electrical instrumentation was dictated by installation of the new torque control drive system.

13. A new hydraulic power unit - equipped with a valve stand - forms the basis of a system that provides positive and accurate control of all moving functions such as rider roll, ejector, core chucks, lowering cradle, and oscillation of the unwind stands. The valve stand, itself, is made up of manifold mounted valves with solenoids possessing indicator lights that show the direction in which the spool of the hydraulic valve has shifted - in other words, direction of oil flow is always in evidence.

Manifold mounting of hydraulic valves permits replacement, or servicing of valves without modification to, or removal of piping. All valves are equipped with gauges to indicate pressure to the pertinent function. Experience proves that systems of this sort facilitate startup of the machine and are time and money savers during periods of trouble shooting.

How much of the original winding system was salvaged, reconditioned, repainted, and reassembled with the new components?

a. Two drums with their bearings and housings.
b. The rider roll assembly with its hydraulically operated rotary control.
c. Two (2) large carrying rolls that were relocated before and after the new Jet-Set slitter section.
d. The main side frames.
e. The hydraulically actuated roll ejector.
f. Most of the unwind and its D.C. drag generator and emergency brake.

<u>Case #2</u>

Following the success experienced with a 217 inch winder, is the renovation of the 266 inch face, heavy duty two drum winder.

This proved to be a notable experience for two good reasons:

1. Because of the magnitude of the rebuild.
2. Because removal of existing parts, replacement with new parts (including the addition of major frame sections), <u>and</u> startup were accomplished over a 6 day shutdown, working two 8 hour shifts per day.

This accomplishment took place at Chesapeake Corporation of Virginia, West Point, Virginia mill during the 4th of July shutdown in 1982. The rebuilt winder operates at 7000 fpm, and is capable of slitting and rewinding 72 inch diameter rolls of 35#/3000 sq. ft. to 47#/1000 sq. ft. Kraft stock.

A condensed list of new components supplied for the rebuild looks like this:

1. Front vertical framework.
2. Wound roll barrier/nip guard.
3. Complete shaftless winding apparatus for 3,4,5 and 6 inch fiber cores.
4. Rider roll with beam and mountings.
5. Rider roll programming.
6. Roller type web holder.
7. Web threading system.
8. Jet-Set^{tm} slitter section sectional rolls.
9. Dual spreader rolls.
10. Electric load cell tension control system.
11. Front drum drive booster.
12. Operators' console.
13. Special coating and grooving of drums.

The size of the wound rolls for which this winder is designed - 72 inches diameter x 258 inches trim - <u>and</u> its top speed of 7000 fpm, dictates massive wound roll barriers, especially on the discharge side! However, the customer wanted to retain his existing wound roll lowering cradle and bridge (or gate). Consequently, because of lack of space, the new barier type nip guard had to be housed within a vertical framwork that, not only, would permit normal raising and lowering functions; but, would comply with applicable safety codes.

This requirement resulted in the addition of vertical framework almost as massive as the existing winder frames. In addition to containing the barrier/guard, this framework becomes the outboard support for the core chucking and rider roll mechanisms.

The bolted-on vertical slideways were removed from each of the existing frames and replaced by shorter bars to which double Roundways are secured. Shortening these bars exposed machined surfaces on the upper portion of each main frame that proved ideal areas against which to fasten the new outboard framework. The new frames were supported partially on new base plates and partially on existing horizontal frame sections between the drum bearings and the wound roll cradle.

The barrier/guard moves smoothly up and down between outboard columns that house its counterweights. Counterweighting is designed to hold the barrier in its momentary vertical location in event of drive failure. A pushbutton operated A.C. gear motor with chain drive is used to raise and lower the barrier.

A drum guard, which protects the operator at threading speed (when the barrier is raised) is designed so that the top section can be manually raised, providing access to the front drum in event paper must be removed due to a web break, or wrap-up.

Since one of the major reasons for rebuilding this winder, also, was elimination of the existing cantilevered coreshaft boxes, slides, raising and lowering and counterbalancing mechanisms - the existing bolted-on coreshaft and rider roll slideways were discarded and anti-friction Roundways installed. With the addition of new outboard vertical frames containing Roundways, the core chucking, <u>and</u> the rider roll mechanisms were restrained between two vertical columns - each end - rather than being cantilevered from a single column, as before. This design, with Roundways on four (4) columns, insures vertical, lateral and longitudinal stability of core chucks, and rider roll.

Through the use of telescoping sections in each core chucking mechanism, trim widths (overall core lengths) from 258 inches to 235 inches can be accommodated while maintaining full stoke of the hydraulic chucking cylinders. This manual adjustment is made through a handwheel, sprocket, chain and screw driver on each chuck. Actual scope of adjustment is approximately 15 inches per side.

Shaftless winding, through the elimination of the coreshaft, allows each individual roll of a set to find its own position in the "winding pocket". It is no longer restrained by the coreshaft, and, as a result, more uniformly would rolls are produced.

Winding cycles are reduced, as is operator fatigue, since coreshafts do not have to be handled. In addition, the maintenance costs associated with coreshafts and their handling mechanisms are eliminated.

Core chucking assemblies are raised and lowered, and counterbalanced, through pneumatic cylinders direct connected to the carriages and to overhead beams.

A compact hydraulic unit, equipped with valve stand, supplies power to the chucks.

A new 10.5 inch diameter, grooved, sectional rider roll replaces the original 8.5 inch single piece roll that had proved too small for the speeds required. New support mechanisms for roll and drive motor are held between the same Roundways that are used for the core chucking mechanism. The existing pneumatic cylinder counterbalancing and pressure meachnisms were rebuilt and all functions programmed for fully automatic, or alternately, manual control. This system allows the operator to set a pre-determined rider roll pressure at the start of the winding cycle, and, also, to select a diameter at which rider roll pressure will be reduced to zero. In the automatic mode, the system will taper the pressure from starting pressure (maximum) to zero at a specific diameter.

Such system assists in building uniform density into the wound rolls as roll diameter builds up, and allows an operator to duplicate previous settings that gave good results.

Where web, or rell conditions dictate, the operator has the option of total manual control during the cycle.

In order to avoid the necessity for rethreading at each set change, a roller type web holder secures the tail between the drums until it can be wrapped around the cores. The drums can be rotated at thread speed and the rubber covered rollers will still maintain contact, holding the tail against the back drum.

A web threading system automatically conveys the tail from the slitter section, through the dual spreaders, and up around the back drum, where it can be attached to the cores. Equipment consists, basically, of air-assist pans and rollers.

A completely new Jet-Set^{tm} slitting station with horizontal web run, mounted independently between existing back frames of the winder, on its own sole plates, replaces the original slitter seciton.

The twenty (20) state-of-the-art units are carriage mounted on anti-friction ways and located in a manner that provides for easy accessibility. Manual positioning is facilitated by a unique system involving locking top and bottom slitter units together - an arrangement that maintains the preset side loading pressure while permitting each set to be repositioned on the Roundways in 15 seconds. This system eliminates 50% of the hoses required by other systems. Slitters are mechanically locked in place during operation and are pneumatically unlocked. Consequently, slitters do not unlock and "float: out of position if mill air supply is lost to any, or all units.

The bottom slitter bands are individually mounted on Fairbanks Morse motors whose supports are designed for quick and easy removal, and the existing top slitters are spring loaded against the bottom bands. The buss bar supplying current in the 460-550 volt range to the slitter motors is uniquely housed between two support beams for the bottom slitters; thus, making a compact, safe, unitized slitter section.

Since the winder is required to rewind a fairly wide range of basis weights in several grades, and to slit a variety of widths in each set, as well as from set to set, a dual spreader system is mounted after the slitters. These grooved, fixed bow spreader rolls function as a set, one roll complementing the other, to uniformly separate and hold slits before the web contacts the first winding drum. Sets are virtually free of stuck rolls, as long as the bowed rolls are in adjustment.

In order to induce the web to flatten and to align itself properly before entering the slitter section, an existing large diameter, smooth face, web driven guide roll was mounted in adjustbale bearings on framework ahead of the slitter station. A modern disc brake replaces the old show type brake.

The rolls located immediately before and after the slitters are small diameter, low inertia, web driven units with sectional aluminum bodies. The diameter of each roll is such that it always operates under the half-critical speed. Vibrations normally experienced at half-critical speed. Vibrations normally experienced at half-critical, which adversely affect web tension sensing systems, are, thus, avoided. This is of extreme importance, especially in this instance, because the roll after the slitters is supported on a set of electric load cells that must accurately sense web tension requirements and transmit resultant information to the controls of the existing unwind regenerative braking motor. The accuracy of this system has a great bearing upon quality of slit and finished sets.

A booster generator was added to the existing M.G. set to increase the torque capabilities of the front drum motor. This additional front drum drive torque, combined with all new drive torque controls, with automatic tension control, and with programmed rider roll pressure, results in control of winding roll density that compares favorably with any new winder.

The operator has virtually unlimited flexibility in his controls. High torque on the front drum, combined with high rider roll pressure at the start of the winding cycle, produces tight wraps around the cores, which is, of course, a prime requisite for winding good rolls. A poor start negates even the most sophisticated controls farther into the winding cycle.

Reduction in front drum torque during the winding cycle assists in controlling wound-in tension produced by nip action between the drums, and by the ever increasing diameter and weight of the winding roll.

These controls function to reduce the amount of internal bursts and potential "starring" of the finished set.

In order to further enhance the effectiveness of these control systems, and, to eliminate the high level of noise created by the existing Chevron grooving, the front drum was resurfaced and tungsten carbide coated, while the back drum was regrooved with a spiral pattern that assists in evacuation of air that winds into the roll with high speed webs. At the time the frums were out of the winder for reworking, a "parking" brake was installed on the back drum - another operator "assist"!

New operator control devices were supplied for the updated electric drive, and for the hydraulic and pneumatic systems. Numerous small panels containing operator controls for specific functions were strategically located on the machine. The main operators' console was custom designed to meet several unique requirements of the customer. These requirements resulted in Appleton's supplying all oeprator devices, the console proper, and the in-factory installation and testing of all components.

Twenty-one Winders With Individually Programmed Stations And Slitters

T. Greiner

As early as 1962, finishing supervisors realized that there were certain limitations to the two-drum winder as the standard piece of equipment for slitting and rewinding paper into rolls. The limitations of the two-drum winder manifest themselves in two areas:

1. The nip pressure between the rewinding rolls and the winder drums is largely dependent on the weight of the rewinding rolls. When this weight reaches a high level, the resulting nip pressure tends to produce excess hardness at the roll periphery and results in over stretching of the sheet.
2. Cross machine profile variations result in defects when each log is wound on one set across the machine, on one pair of drums under one rider roll. Common defects are ridges, welts, and stuck rolls.

Relief from these problems has been sought and partially achieved through the use of large diameter drums, variation of torque between the winder drums, and shaftless winding. Problems have continued with two-drum winders, however, especially in the processing of high density coated papers, such as LWC and heavily filled sheets. These sheets can have an apparent density of 1.25 g/cm^3, which is 25% greater than that of water, as compared to uncoated papers, which have densities of about 0.7 to 0.8-g/cm^3. Problems with high density sheets became particularly acute when printers required rolls with diameters exceeding the 1000 to 1100mm (40 in.-44 in.) range.

DEVELOPMENT OF THE INDIVIDUAL STATION WINDER

These problems gave birth to the MIR Winder, the first single drum winder with individual stations for each finished roll which are staggered againest one rewind drum to provide surface drive (Fig. 1). With the development of the new winder, it became possible to control the nip pressure of each rewinding roll againhst the drum, regardless of roll weight. The surface drive is assisted by by applying additional torque to the rewinding cores.

1. *Original MIR Winder.*

Further development of the individual station principle resulted in the Biwind Winder (Fig. 2). The hydraulic center drives were dispensed with and the single drum was replaced with two vertically mounted drums. The chuck bearings move along sloping paths, so that the wind is started in the nip between th two drums, but continues againest a single drum after the rolls have reached a certain diameter.

2. *BI-WIND Winder.*

A similar idea is incorporated in the Vari-Roll Winder (Fig. 3). It uses a single-large diameter drum with twin rider rolls to provide a tight start on rolls exceeding a certain width, where core deflection can become a problem.

3. *VARI-ROLL Winder.*

On this winder, the twin rider rolls are pressed against the starting rewind roll by an air cylinder. The rider rolls are lifted off automatically when the rewinding rolls reach a programmed diameter.

PROS AND CONS OF INDIVIDUAL STATION WINDERS

All of the above individual station winders have wound very good rolls under marginal to impossible conditions for two-drum winders. High density coated and filled papers, as well as specialty products like glassine, can now be wound to diameters of 1270mm (50") or more with uniform roll density. Cross machine profile variations are not nearly as much of a problem on these winders, since each finished roll can assume a different diameter, and roll separation is guaranteed. A splice can be made on an individual roll without affecting the remainder of the set.

There is, however, a price that must be paid for all these benefits, and I am not referring to the cost of the equipment, but to the downtime inherent in the operation of this class of winder. Since the winder must follow a paper machine, an off-machine coater or a super-calender, its speed and duty cycles must be so designed that it will not become a bottleneck to the production unit ahead of it. It stands to reason that more time is required to core up individual stations than to insert all the cores of a set into a two-drum winder. Short rums in particular can be a problem with an individual station winder, as the two frames supporting every rewinding roll must be reset to the order pattern before the roll is run. Slitter setting on these machines is not nearly as fast as on two-drum winders because the sheet run necessitates mounting the slitters high on the machine (Fig. 1).

DOWNTIME IS THE ENEMY

Winding is a batch operation, in contrast to production on a paper machine or a coater. Two types of downtime occur on every order:

1. Change time, rewind; this is the time required to discharge a set and start a new one from the same reel.

2. Change time, rewind + reel; as above, but including a new mill roll and rethreading the winder.

Change time at the rewind depends upon the width of the winder and the number of stations engaged. Change time can be reduced through proper personnel assignment and organization, as well as by devices which aid in coring and roll starting. The additional time for the reel change is influenced not so much by the width of the machine as by the method of threading (which should be largely automatic) and organization (availability of crane, etc.)

There is, of course, a third category of downtime, namely, that required for order changes, splicing, knife changes, maintenance, etc. The time required for this category must be estimated according to the order pattern at the particular machine and the degree of automation of the winder. When estimating winder productivity, we will refer to this third category of downtime as "available time", and assign a specific number of minutes in each shift for it.

HOW TO GAUGE PRODUCTIVITY

We will now turn to practical examples of estimating our time to assure that our winder keeps up with a paper machine running at a given speed. We will assume, for the purpose of this exercise, that the paper machine operates at 650m/min (2132 fpm), and that we want one hour of available time in every 8-hour shift.

The parameters we require for this calculation are the footage or lineal web length of paper wound on a finished roll, rate of acceleration number of sets per reel, and required change times at the rewind and unwind ends. The footage on each roll is calculated from the diameter, the caliper of the paper, and the OD of the core. Given these parameters, a computer can print out a family of curves for given amounts of available time. In these curves, paper machine speeds are plotted againest winder speed requirements.

Our first example, Fig. 4, assumes that the paper has a caliper of .003, rolls are 40 inches in diameter, that each roll is built for three sets; it also allows 2.5 minutes for set changes and 5.0 minutes to load and thread a new reel. Under these conditions, the winder must be operated at 1040/m/min or about 3400 fpm to keep pace with the paper machine. No doubt, this is an easy order to run.

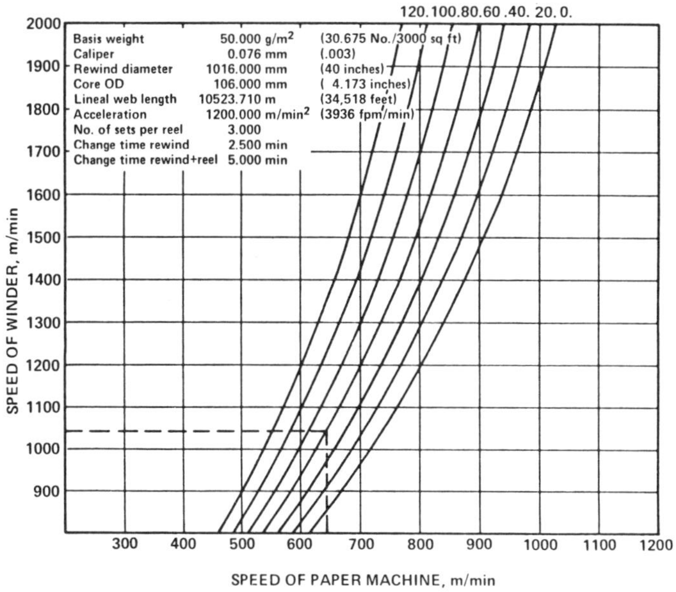

4. *Available time/8 hr. shift.*

Fig. 5 shows somewhat different parameters. While the caliper remains about the same, roll diameter is somewhat higher, which is in the winder's favor, since it results in almost 12 percent more footage per roll. This time, however, only two sets can be made from every reel, so that every other set change means a reel change as well. Moreover, this winder is wider and requires 3.5 minutes per change at the rewind and 7.0 minutes when the reel change is included. The result is that this winder must be operated at 1250m/min or about 4100 fpm to keep pace with the paper machine.

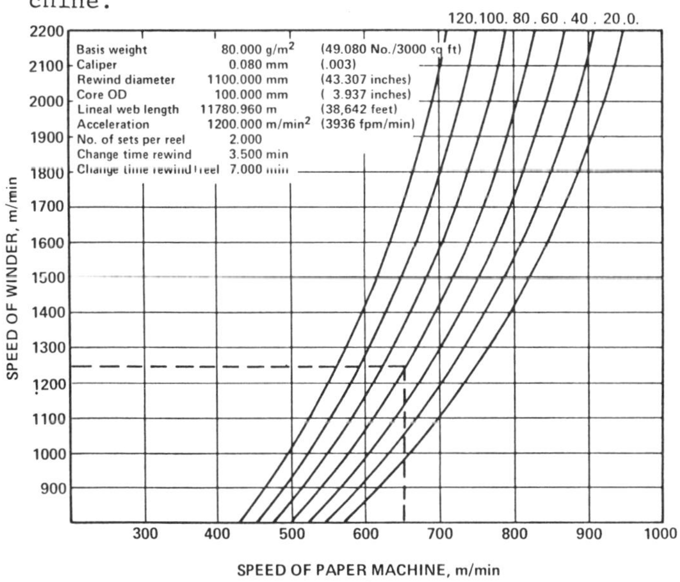

5. Available time/8-hour shift

Our last example, Fig. 6, shows a heavier sheet with almost 22 percent less paper on each roll, even though the diameter is greater. To make matters worse, each reel holds only one set, so that each set change requires a reel change as well. 6.0 minutes are programmed for each set change. You can see that the winder must now be operated at about 1930 m/min or 6336 fpm to keep up with the same paper machine, if the 60 minute "available time" requirement for order changes, maintenance etc., for each shift is to be met.

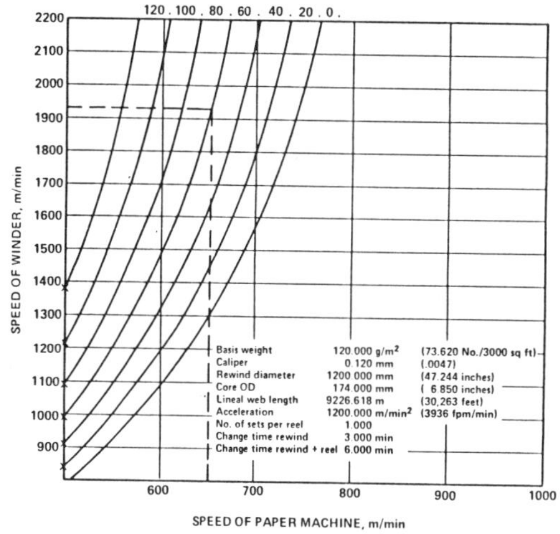

6. Available time/8-hour shift

PROGRAMMING TO ELIMINATE SETUP TIME

Having used up all means of keeping change time as short as possible, we must turn our attention to the time consumed for order changes. On a two-drum winder, this consists primarily of slitter setting, a procedure that normally requires one to two minutes per slitter, depending on its design. On a winder with individual rewind stations, this operation is more complex. While the slitters themselves are similar to those on two-drum winders, their location is rather awkward, as dictated by the sheet run on this type of winder.

In addition to slitters, however, we must set a separate rewind station for each finished roll, i.e., two supporting frames per roll. If we want to wind a set of eight rolls, we must set nine pairs of slitters plus sixteen roll frames, and, sometimes, a few rider rolls as well. It stands to reason that automation through electronic programming can play a vital role in this area.

The mechanical premises must be created first. Rack and pinion movement is generally considered the most precise and reliable method of shifting the slitters. Ball bushings are frequently used to reduce friction and eliminate possible jamming. Top and bottom slitters must be aligned correctly vis-a-vis each other. Each top slitter can be connected mechanically to its bottom slitter for the purpose of travel. In this manner, only one motor is required for each pair of slitters and correct alignment is guaranteed. While this requires the web to be broken during slitter setup changes, it is normally not considered a drawback, because such changes usually coincide with reel changes. The alternative requires a separate motor for every top and every bottom slitter, resulting in potential synchronization and alignment problems.

Motorized remote-controlled movement of each station requires similar mechanical considerations, and poses additional problems. The inertia of each supporting frame is greater than that of the slitters, but it is vital that each station assumes its exact positions. If cores of non-uniform diameters are used (especially iron cores) there is a danger of incomplete chuck insertion. Monitors are needed to advise operators that not only have all elements assumed their assigned positions, but that all cores have been properly inserted.

The use of electronics in general and process controlling equipment in particular has made it possible to program every slitter and station element in advance, i.e. while a previous order is being processed. The most commonly used equipment employs a command console with a keyboard and LED displays to show the operator the current as well as the newly programmed position of every element.

MICROPROCESSOR CONTROLS

More advanced systems are available today. Shifting motors are controlled by so-called absolute encoders, rotary postion sensors which can operate with a resolution of 0.25 mm (0.10 inches). This means that they can signal accurately the position of each element within these tiny steps. It stands to reason that mechanical tolerances must be taken into consideration as well, so practical accuracy will be somewhat less. Nevertheless, it will be at least as great as that possible when manual setting is used.

Moreover, a system with such fine resolution and the means of reporting each position enables the operator to decide whether the deviation displayed is acceptable or whether a manual correction is indicated.

Once a microcomputor system with a video screen monitor is used, there are further refinements which can be of tremendous aid to the operator. The microcomputor can be programmed to show not only the setting of every position and the actual status, on the screen can also flash a warning when a particular element has not reached its assigned position, and indicate the reason for the malfunction. It can be programmed also to monitor its own electrical circuits and identify the source of any possible failure. Then, it can play big brother to the operator and make sure that he does not punch in a program which is not feasible. For example, it will refuse to accept a program that exceeds the width of the machine of that includes rolls wider or narrower than the mechanical limitations of the winder permit.

The important thing is that all this takes place while the winder is operating and therefore does not produce any downtime. The only downtime is that required for the elements to position themselves automatically in accordance with the prepared program. This time is very short, ranging from a few seconds to a minute or two, depending on the system, width of the machine, number of stations, magnitude of order change, etc. Since this time generally coincides with loading of a new reel, there is practically no time lost for the operation.

OPERATION FROM A CENTRAL COMPUTOR

The next and final step is that of connecting the programming to a remote source, such as a mill's scheduling computor. A variety of possibilities arise once we enter this phase. Punch cards can be prepared at a mill's central scheduling facility and fed into the winder's programmer, thereby relieving the operator of programming the run. The programmer can also be directly connected to the mill's computor so that the operator can call up the setup by order number and the program will be prepared without his assistance.

While such programming can be used for slitter setting of all types of winders, its benefits are greatest on winders that require above average time for order changes. Programming has eliminated the drawback on winders which have given outstanding quality performance on hard-to-wind grades.

Twenty-two
A Winder For Heavy Rolls

Thomas S. Greiner

Management Policy Adviser, Jagenberg, 645 Madison Ave., New York, N.Y. 10022

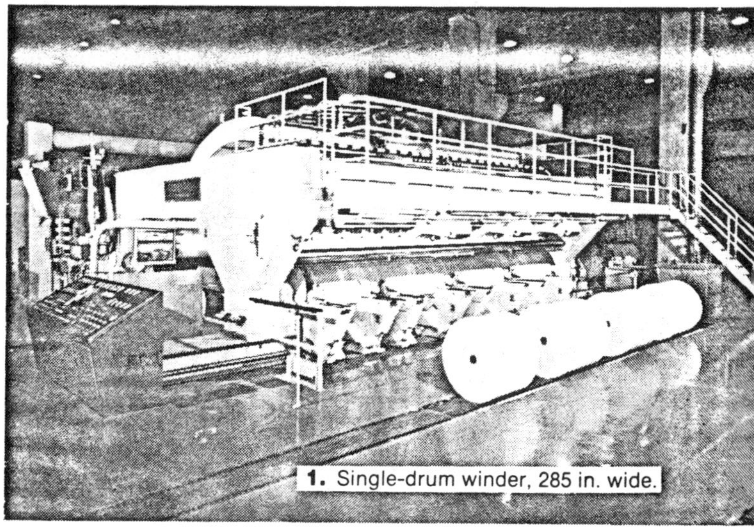

1. Single-drum winder, 285 in. wide.

Winders with individual rewind stations, such as single-drum winders, have proven themselves the preferred means to wind high-density papers to rolls of uniform structure; this is true especially of coated and other high-finish grades. These winders feature shaftless windups for each finished roll. Such winders are limited, however, by the maximum weight which can be supported by a pair of chucks and by the core of the roll. Modern rotogravure presses often require rolls of considerable width and diameter and use papers of high apparent density. To the weight of these rolls must be added the nip pressure required for sufficient contact with the carrier drum for surface driving and control The need to wind such heavy rolls on individual stations has led to the development of the Vari-Top* winder, which supports part of the weight of the rewinding rolls on the center drum and operates at pre-selected nip pressures by means of programmed load-relief mechanisms. Individual rolls of good structure can be wound in this manner, even if their weight considerably exceeds the limits of the conventional single-drum winder.

Keywords: Winders • Finishing • Equipment • Engineering

The single-drum winder with individual rewind stations has taken over on applications which exceeded the limitations of the conventional 2-drum winder. The demand for rolls of uniform and controlled hardness on rolls with diameters of more than 40 in. (100 cm) on papers with high apparent density has forced departures from the single 2-drum winder, which had been the standard piece of equipment in the industry from the beginning. The need for individual treatment of the rewinding rolls becomes even more pronounced when cross-machine profile problems occur, which are especially difficult to control on wide paper machines.

---

*Trade name.

The single-drum winder has attained maturity

The main impediment of the single-drum winder as a piece of production equipment had been the downtime required to make ready each rewind station. This has been overcome by programmed and automated slitter and station settings. Threading aids have been devised to ease operation and to shorten downtime on reel changes. Today's automated single-drum winder can keep up with high-speed paper machines and off-machine coaters on a one-on-one basis in practically all applications (*1*). Figure 1 is a typical example of such a winder as major production equipment.

As long as individual roll weights are not extremely high, winders of this design are used very successfully on even the widest paper machines and off-machine coaters. This is true quite particularly on offset grades of coated papers. Rolls for offset printing are rarely over 40 in. wide. Papers made for web offset have good strength, low moisture content, and little, if any, fillers. Their density is thus not unreasonably high.

Limitations of single-drum winders

Under certain conditions, however, the conventional single-drum winder has reached its limitations. Figure 2 shows a schematic of this method of winding. There is no support for the individual rewinding rolls except through the chucks which carry each roll on its core. The nip contact with the single-winder drum does not relieve the load on the core but adds to it because nip pressure is required to transmit torque and to control roll structure.

This situation becomes critical when rotogravure printing is considered. Although this type of printing is not as dominant in this country as it is in Europe, it is gaining ground and is used today not just for Sunday supplements and catalogs but also for major publications printed on coated papers, like *National Geographic* or *Playboy*.

The roll width problem

Since rotogravure presses are built for high production, many of them are very wide. They can take wide rolls, because they do not have the register problems encountered by offset printers on wide webs. Rolls of 70–94-in. width are quite normal. The widest rotogravure press presently in use can run rolls 110 in. wide (*2*). It is not surprising that printers who operate such units demand as much footage on a roll as possible to avoid an excessive number of pasters. This translates to larger diameters and, consequently, to yet higher roll weights. Since weight of rolls grows with the square of the diameter, this factor becomes even more critical than width when computing roll weight.

The paper density problem

Finally, we must consider the physical properties of the paper itself. The rotogravure printing process demands the smoothest surface and the most flexible sheet possible. Supercalendering will provide the surface. High filler content will make the sheet flexible and also aid in achieving opacity, which becomes highly important as basis weights are reduced to counter the high cost of mailing.

Papers for rotogravure printing also tend to run to higher moisture levels than those for web offset, again in an effort to achieve a flexible sheet. All those factors add up to extraordinarily high density, i.e., ratio of weight to volume. Lightweight coated papers (LWC) for rotogravure printing may have an apparent density of 1.25 g/cm^3 or 0.045 lb/in.^{3} This means that this paper is 25% heavier than water or about 50% denser than most offset printing papers.

The roll burst problem

Conditions such as those described above lead to overstressing of the paper, resulting in bursts. They especially occur near the center of the roll and are caused by excessive nip pressures at the end and softness of the roll at the center, as demonstrated by Hoelzl and Wiesse (*3*). The combined effect of roll width and roll diameter, as well as high density and low strength of the paper, all contribute to produce this problem.

The core problem

One way to alleviate these conditions would be to increase the diameter of the cores from 3 in. ID to, say, 6 in. ID. The larger core would give the roll more structural rigidity and increase the area of contact with the winder drum, thereby reducing linear pressure. While some efforts are being made in this direction, there is little hope that a change will be made in the foreseeable future, since printers oppose larger cores because practically all the unwinds at the printing presses are equipped to handle 3-in.-ID cores. Switching to another diameter entails not only a change in unwinding chucks and adjustment of diameter-sensing equipment, but it also means handling heavier cores. Even more critical may be the danger of damage in transit, since large-diameter cores are less resistant to crushing than small-diameter ones.

Most American gravure printers have a preference for returnable iron cores. These have the advantage of greater strength but also, most importantly for the printer, of crush resistance. Unfortunately, the Standard 40 iron pipes used for this purpose have sloppy tolerances in OD and ID and are frequently bent, especially after a few round trips. Straightening devices are used but are only partially successful. Europeans have experimented with plastic inserts to reinforce regular paper cores. To date, however, the most suitable cores seem to be extra thick, high-quality fiber cores.

Center-wind assists are one approach to the problem. The idea is to strengthen the wind at the start of the roll without using excessive nip pressure. Applied research conducted at one German paper mill has not led to a satisfactory solution along this line. It did not appear possible to apply a sufficient amount of torque to the core without seriously damaging the core itself. If this was to be avoided, however, the effect of the center-wind assist did not prove sufficient to prevent roll-center bursts on the types of rolls described earlier.

The Vari-Top

The solution was found in changing the configuration of the single-drum winder to shift the line of contact between the rewinding rolls and the winder drum to an angle of approximately 30° from the vertical, as illustrated schematically in Fig. 3. In this manner, a portion of the growing roll weight of the winding rolls is absorbed by the carrier drum. The nip pressures between the rolls and the drum are freely selectable, because the chuck assemblies for the station can be loaded and counterloaded with infinite variations. Nip pressure is programmed in accordance with the requirements of a specific grade of paper and can be as little as 4 kg/cm (22.4 pli) on rolls of 50-in. diameter.

The diagrams in Fig. 4 demonstrate the difference in resulting nip loads on the cores of the Vari-Top, contrasted with those on a regular single-drum winder. The core loading (CL) on the single-drum winder is equal to the hypotenuse of a right-angle triangle, of which roll weight

2. Schematic of single-drum winder.

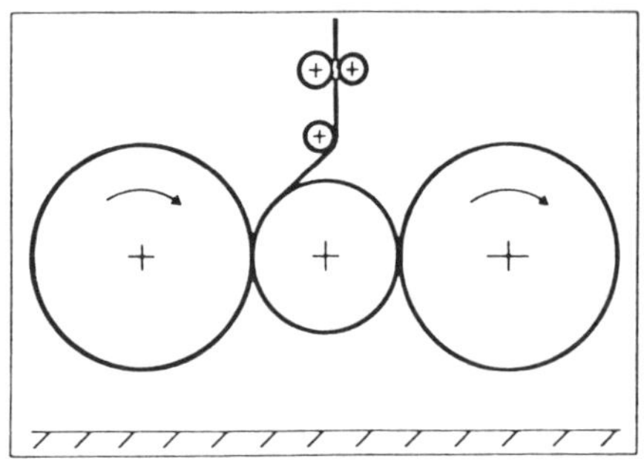

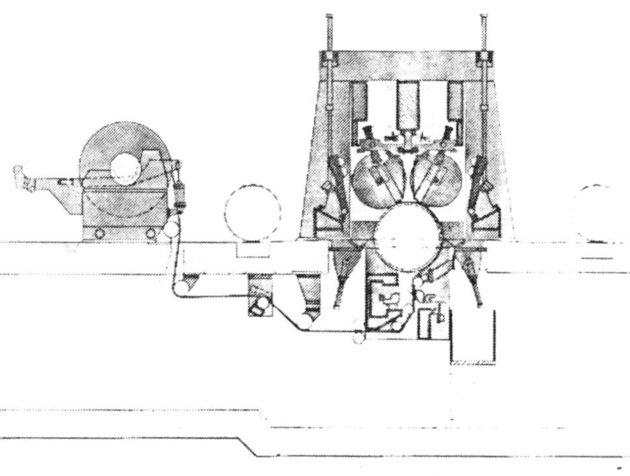

3. Schematic of Vari-Top.

(W) and applied nip pressure (N) form the short sides. Now, if we change that right angle to an acute one and apply the same nip pressure, the core load becomes considerably smaller, as shown in the diagram of force vectors labeled "Vari-Top." Vector E represents the cylinder force for load relief, and Vector F, the force absorbed by the gibs of the chuck carriages. Looking now at the curves, they show how the core load increases with increasing roll diameter. Curve W shows the nip load on the core without applied pressure. The upper three curves represent progressive values of core loads when the applied nip load is 4, 6, and 8 kg/cm, respectively, equalling approximately 22, 34, and 45 pli.

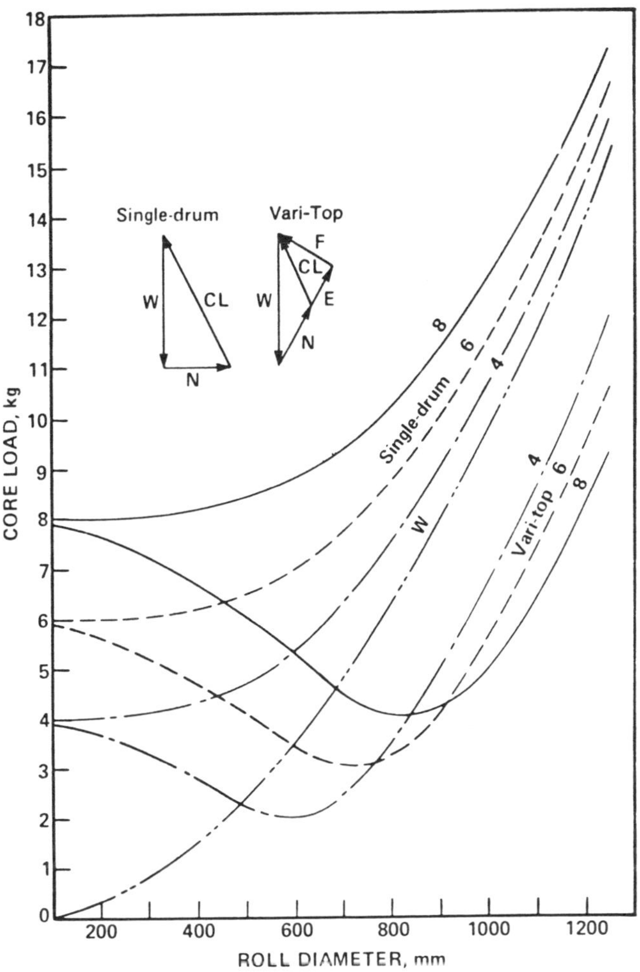

4. Core loading vs. roll diameter on regular single-drum winder and on Vari-Top at roll width of 1 cm, apparent density of 1.25 g/cm³, and applied nip pressures of 4, 6, and 8 kg/cm.

A practical example calculated from the values of these curves will illustrate the problem and its solution. If we assume that a roll of LWC paper for rotogravure has a width of 230 cm (90.5 in.) and a diameter of 125 cm (49 in.), we arrive at a roll weight of approximately 3500 kg (7700 lb). Adding to that the necessary nip pressure will result in a core load of about 3700 kg (8160 lb) or more than even the best quality core can tolerate.

The lower three curves are those resulting from the geometry of the Vari-Top. Using the above example, if we apply full weight relief, the total nip pressure against the drum would be about 920 kg (2030 lb). The geometric load on the core would calculate to about 2800 kg (7110 lb). Experience has shown that roll bursts can be avoided under these conditions.

Figure 5 is a photograph of the windup section of the Vari-Top showing the rewind stations and their load-relief mechanism. As previously mentioned, the nip pressure is programmed and can be selected for five different conditions, depending on the type of paper and width of rolls. The minimum nip pressure at 127 cm (50 in.) diameter is about 4 kg/cm (22 pli). The carrier drum has a diameter of 1500 mm (59 in.). The 30° angle of the rewind stations from the vertical has the added advantage that the difference in wrap angles between the front and back stations is very small, resulting in almost equal conditions.

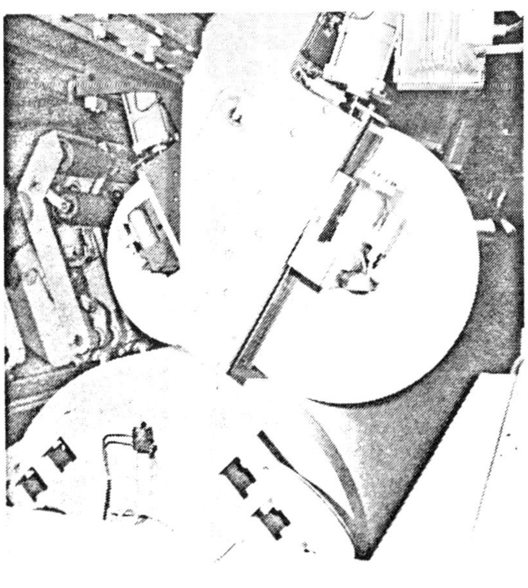

5. Staggered rewind stations on Vari-Top.

Field experience

The first Vari-Top has been in operation at MD Paper Mills in Plattling, Germany, since the summer of 1982. It is 580 cm (228 in.) wide and operates behind a paper machine producing LWC papers. The following are regular production data on tested runs of 250 rolls in that mill:

Roll width, 2316 mm (91 in.)
Roll diameter, 1100 mm (44 in.)
Core ID, 76 mm (3 in.)
Core OD, 124 mm (4.875 in.)
Core inserts, 350 mm (13.75 in.)
Winder speed, 1600 m/min (5200 ft/min).

Test rolls of 2640-m (104 in.) width and 1250-mm (49 in.) diameter have been run. There were no bursts in the paper of any of the rolls produced on this system.

Conclusion

Heavy and dense rolls, such as those frequently required in LWC papers for rotogravure, cannot be wound satisfactorily to large diameters on regular single-drum winders. This limitation can be overcome by relocating the rewind stations to the vicinity of the apex of a large-diameter carrier drum and by carefully controlling nip loadings. The equipment designed for this purpose has given superior results under realistic operating conditions.

Literature cited

1. Greiner, T.S., *Tappi J.* **63**(10): 39(1980).
2. Wilken, E., and Ynostroza, R., The Facts on Gravure, *Graphic Arts Monthly* March: 52(1983).
3. Hoelzl, E. and Wiesse, J., *1982 Paper Finishing and Converting Conference Proceedings*, TAPPI PRESS, Atlanta, p. 111.

Received for review July 22, 1983.

Accepted Dec. 29, 1983.

Based on a paper in *1983 Finishing and Converting Conference Proceedings*, a TAPPI PRESS publication.

Twenty-three
Taking Core Manufacture Out Of The Dark Ages

D. Johnson

1. Paper Mill Needs
   As they are equipped with different types of winders, the individual production units of Escanaba have different core requirements.

   a. No.1 Mill (single drum winder)
   b. No.3 Mill (single drum winders with torque drives)
   c. No.4 Mill (2-drum winders)
   d. Reprocessing winders (2-drum)

2. Fiber and Iron Cores
   a. General description of each type.

3. Fiber Cores
   a. How we select material
   b. Transport - its problems
   c. Handling and storage
   d. Identification of cores
   e. Metal caps and sleeves
   f. Core processing and equipment
   g. Quality control.

4. Iron Cores
   a. How we select material
   b. Transport - its problems
   c. Handling and storage
   d. Core processing and equipment
   e. Quality control.

5. Iron vs. Fiber Cores - Pros and Cons
   a. Handling, how does it differ?
   b. Storage, what to be aware of
   c. Comparing weights
   d. Accountability of returnable cores
   e. Runnability on the winder

6. Our commitment to Core Processing
   a. Old vs. new facilities
   b. Advantages and disadvantages

CONVERSION FACTORS FOR SI UNITS

Quantity or Test	Value in Trade or Customary Unit	X	Conversion Factor	=	Value in SI Unit	Symbol
Area	square inches		6.45		square centimeters	cm^2
	square feet		0.0929		square meters	m^2
	square yards		0.836		square meters	m^2
	acres		0.405		hectares	ha
Basis Weight* or Substance (500-sheet ream) or Grammage* when expressed in g/m^2	lb (17 x 22 - 500)		3.760		grams per square meter	g/m^2
	lb (24 x 36 - 500)		1.627		" " " "	"
	lb (25 x 38 - 500)		1.480		" " " "	"
	lb (25 x 40 - 500)		1.406		" " " "	"
	pounds per 1000 square feet (Paperboard)		4.882		" " " "	"
Breaking Length	meters		0.001		kilometers	km
Burst Index	$\frac{g/cm^2}{g/m^2}$		0.0981		$\frac{kPa}{g/m^2}$	
Bursting Strength	pounds per square inch		6.89		kilopascals	kPa
Caliper	mils		0.0254		millimeters	mm
Concora Crush	pounds		4.45		newtons	N
Edge Crush	pounds per inch		0.175		kilonewtons per meter	kN/m
Energy	British thermal units (Btu)		1055		joules	J
Flat Crush	pounds per square inch		6.89		kilopascals	kPa
Force	kilograms		9.81		newtons	N
	pounds		4.45		newtons	N
Length	angstroms		0.1		nanometers	nm
	microns		1		micrometers	µm
	mils		0.0254		millimeters	mm
	feet		0.305		meters	m
Mass	tons (2000 lb)		0.907		metric tons	t
	pounds		0.454		kilograms	kg
	ounces (avd p)		28.3		grams	g
Mass per Unit Volume	ounces per gallon		7.49		kilograms per cubic meter	kg/m^3
	pounds per cubic foot		16.0		kilograms per cubic meter	kg/m^3
Puncture Resistance	foot pounds		1.36		joules	J
Ring Crush	pounds (for a 6-in length)		0.0292		kilonewtons per meter	kN
Stiffness (Taber)	gram centimeters (Taber Units)		0.0981		millinewton meters	mN.m
Tear Strength	grams		9.81		millinewtons	mN
Tensile Breaking Load	pounds per inch		0.175		kilonewtons per meter	kN/m
	kilograms per 15 millimeters		0.654		kilonewtons per meter	kN/m
Volume, Fluid	ounces (US fluid)		29.6		milliliters	mL
	gallons		3.79		liters	L
Volume, Solid	cubic inches		16.4		cubic centimeters	cm^3
	cubic feet		0.0283		cubic meters	m^3
	cubic yards		0.765		cubic meters	m^3

*See TAPPI Technical Information Sheet 018-4.